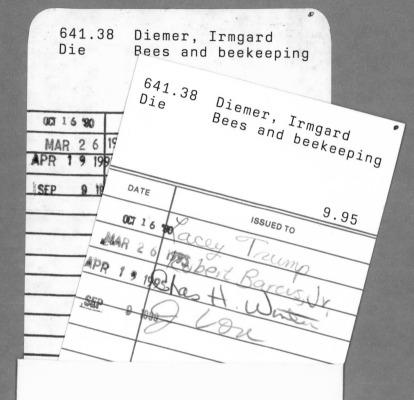

641.38   Diemer, Irmgard
Die   Bees and beekeeping

OCT 16 90
MAR 2 6 19
APR 1 9 199
SEP   9 19

641.38   Diemer, Irmgard
Die   Bees and beekeeping

9.95

DATE	ISSUED TO
OCT 16 90	Lacey Trump
MAR 2 6 19	Robert Barcus Jr.
APR 1 9 1995	Chas H. Winter
SEP 9 1998	J. Love

BEES and BEEKEEPING

BEES and BEEKEEPING

Irmgard Diemer

MEREHURST PRESS
LONDON

Cover photo: WBC Beehive in a Meadow, by Martin Dohrn (Science Photo Library)

Published 1988 by Merehurst Press
5 Great James Street
London WC1N 3DA

Co-published in Australia and New Zealand by
Child & Associates
5 Skyline Place
Frenchs Forest
NSW 2086
Australia

First published in German in 1986 under the title *Bienen*.
German edition © 1986 Franckh'sche Verlagshandlung W. Keller & Co. Kosmos-Verlag, Stuttgart, Federal Republic of Germany

Picture credits
Drawings by Linda Waters.
Photographs by Geoffrey Lawes except pages 36, 40 and 45 (Stephen Dalton/NHPA) and page 133 (Martin Birley/Tropix Photographic Library).

Editor Lesley Young
Translated by Elizabeth Doyle in association with First Edition
Designed by Carole Perks
Typesetting by Maureen Tunningley
Reprographics by Fotographics Ltd, London-Hong Kong
Printed in Portugal by Printer Portuguesa Industria Grafica LDA

CONTENTS

Publisher's Note

Where specific months are mentioned instead of seasons, these are broadly appropriate for the seasonal conditions prevailing in the temperate climate of such northern hemisphere countries as Germany, France, Britain or New England. Even within a small region climatic variations can lead to significant differences in the yearly cycles of honeybee colonies, so that readers living in other climatic zones must make adjustments accordingly.

Adaptations to the main text to take account of both German and international beekeeping practices (particularly pages 53-67) are by Geoffrey Lawes.

INTRODUCTION

Why keep bees?

If you ask a beekeeper this question, he will probably look at you quizzically and say: 'Bees? I keep them because I like them!'

This is certainly as good a reason as any for keeping bees, but the inner satisfaction for many beekeepers comes in part from their recognition of the vital role played by bees in nature.

In ancient times the bee was a symbol for priests and kings. Throughout history it was considered a sacred creature, with its products held in high esteem.

The ancient Egyptians, for instance, presented honey and wax on the signing of peace treaties – in comparison slaves were regarded as second-class gifts. In the third century BC, the bee was the emblem used on coins in the Greek city of Ephesus and was the symbol of the goddess Artemis.

But it was not only the ancient Egyptians and Greeks who honoured the honeybee; it features widely in the mythology of many countries throughout the world.

The Germanic tribes, for instance, depicted the top of their tree of life, Yggdrasil, as surrounded by bees. Obviously they were aware of the prime importance of bees in the life-cycle of many plants.

The French Emperor Napoleon

The well-known wrought iron gate at the National Agricultural Centre, Warwickshire, England, displays the characteristic shape of a queen bee

was also a bee enthusiast, and had his imperial robes embroidered with golden bees.

Even today bees fascinate many people because of their ordered social structure in which thousands of bees can live together harmoniously without robbing or killing each other. They feed on pollen and nectar from plants; products which would otherwise go unused.

The amazing fact is that in taking, they give something back as well; giving insect-pollinated plants the chance of propagation by transferring pollen from one to another and enabling seed formation.

The Indirect Value of Bees

Bees account for 80 per cent of all pollination by insects. Honeybees alone carry out more pollination than butterflies, wasps, bumblebees and flies put together.

They have the advantage of being able to pollinate many different types of plant rather than being restricted to just a few, as many wild bees are. Because they overwinter in a colony, large numbers of honeybees are ready to collect pollen in the spring, when most plants are in bloom. With bumblebees, however, only the females survive the winter and they then have to rear offspring. In addition, honeybees practise flower fidelity and do not pollinate at random, which ensures that the pollen reaches the correct plant.

Recent research confirms the value of honeybees in pollinating a wide variety of plants including sunflowers, oil-seed rape, clover, alfalfa, cotton and fruit. In the process, the bee influences not only the quantity of fruit or seeds produced, but also their quality.

In some cases a single visit by a bee may not be enough. In one experiment with strawberries, for example, bees alighting one to fifteen times per blossom achieved 100 per cent pollination, but the average weight of fruit of each strawberry was only 5.2 g, compared with 8.1 g where the plants were visited between twenty-one and twenty-five times.

In the South of France an experiment in lavender fields showed that while the flowers on plants pollinated by bees ceased blooming sooner, they also yielded 16-20 per cent more lavender essence.

In another experiment, alfalfa not pollinated by bees produced a yield of 57 kg (125 lb) of seed per hectare, compared with a yield of 200 kg (441 lb) where the crop was pollinated by bees alone, and 232 kg (511 lb) following pollination by a variety of insects. In Germany, pear trees produced only 45 kg (99 lb) of fruit without bee pollination, but 156 kg (344 lb) with bee pollination. Oil-seed rape can yield 53 per cent more oil when visited by bees. In this case pollination increases the number of grains, the number of seeds per husk and the weight of seed; it also improves germination.

Animals also depend on bees. Many kinds of birds live off seeds, fruits and berries which can only grow if bees have previously pollinated the plants.

The economic value of the bee is therefore considerably greater than that of a mere producer of honey. Bees hold a key position in the natural order and are of inestimable importance to animals and plants. They are also of direct benefit to humans who, for centuries, have used their products, including honey, propolis and bee venom, as food and medicine.

The Direct Value of Bees

From earliest times **honey** has been regarded as priceless – food for the gods, in fact. To get some idea of its value, just think of its delicate raw materials and complex production process. It is both a food and a medicine. Because of its high, easily digestible sugar content, it is an ideal source of energy for both the healthy and the sick.

It is absorbed directly into the bloodstream without any further digestion, where it has a beneficial effect on the heart and nerves. It can increase the haemoglobin content of the blood, which is why invalids, especially children, often used to be given treatments of milk and honey.

It can also improve the physical and mental wellbeing of older people who may have problems with their digestion.

Beeswax forms the 'scaffolding' of the bee colony. It is produced by bees during the process of transforming nectar into honey and is secreted from glands in the bee's abdomen. It comprises some 72 per cent esters and 14-15 per cent free ceric acids and has a melting point of around 63°C (145°F). Beeswax has numerous uses in drug and cosmetics production, in artists' materials and in the production of furniture polish and grafting wax for trees.

Like honey, beeswax has been highly prized since earliest times, and for centuries was one of the few sources of artificial light. Even today beeswax candles are in great demand for their warm light and fine scent.

Pollen While many plants rely on bees to enable them to form seeds and reproduce themselves, the bees, for their part, need pollen in order to be able to produce their offspring. Every bee colony needs an average of 30 kg (66 lb) of pollen per year. Each minutely structured grain of pollen contains nitrogenated compounds, carbohydrates, fatty substances, enzymes, minerals and vitamins.

Pollen is also an important health food for humans, although it is often difficult to digest. Since all honey contains a certain proportion of pollen, it is not usually necessary to take any extra. For chronic illnesses, however, a course of pollen treatment can often provide an effective cure.

Pollen is said to stimulate the appetite, improve the metabolism, regularize bowel functions, improve vision and help with prostate conditions.

Worker bee storing 'bee bread' (compacted pollen)

Lumps of propolis that the bees have built at the hive entrance as a defence

Pollen prepared and matured in the comb becomes 'bee bread'. The bees push the pollen down firmly into the cells of the honeycomb with their heads and fill the cells with honey. Enzymes then cause lactic fermentation, breaking down and conserving all the easily digestible components. This fermented pollen is of greater nutritional value for humans than freshly collected pollen.

Propolis or bee glue, is the reddish resin which the bees use to cover walls, frames, honeycomb cells and any cracks in their hives. Any intruders, such as mice, which get into the hive are stung to death and then embalmed with propolis to prevent them from decomposing.

The bees collect the propolis from the buds of trees such as alder, birch, horse chestnut, poplar, fir, pine and cherry trees. The sticky, antibiotic resin is mixed by the bees with wax and oily pollen balsam which they produce through digestion of the pollen spores. Bee glue is made up of roughly 55 per cent resins and balsams, 30 per cent wax, 10 per cent essential oils and 5 per cent pollen, although the actual components vary according to where they come from.

Propolis has a wide variety of uses. It has been successfully used to disinfect and heal wounds and to treat corns, receding gums and diseases of the upper respiratory tract. It has also been used to varnish violins to give them greater resonance.

Royal jelly is the bitter-tasting substance produced by the worker bees to feed the queen. It is obtained from swarm and queen cells. Despite its high price, it is much sought after because of its reputed rejuvenating qualities.

Like snake venom, **bee venom** has been used throughout centuries for medicinal purposes. It could well be said that the bee invented medicine and the hypodermic injection long before humans. Bee venom is said to alleviate and sometimes even cure rheumatic complaints, arthritis, neuritis, neuralgia, bronchial asthma, migraine, inflammation and high blood pressure.

Even cholesterol levels can be controlled with the aid of bee venom. In Russia, living bees are placed directly onto the parts of the patient's body where the venom is to be injected.

The important point to remember is that a bee sting may make you feel drowsy and faint at first, but this is a natural reaction. Generally speaking, the human body adapts quickly to bee venom and parts which have been stung repeatedly eventually cease to react.

Many beekeepers believe that bee venom improves the body's resistance to illness. This could explain why people who have kept bees for years often remain robustly healthy well into old age.

Before the Second World War bees were kept on almost every farm, but with the improvement of economic conditions and the advent of modern farming methods, we seem largely to have forgotten the value of the bee. Over the course of time, bees have become dependent for their very existence on the help and goodwill of man, so it is essential that we humans start paying more attention to bees again and that more people take up beekeeping.

Keeping bees is a restful and relaxing occupation which can broaden your horizons and make you more aware of nature. The beekeeper has an ideal opportunity to learn about plant life and to see how plants change with the weather and time of year. In the course of time the beekeeper can develop an insight into the patterns of nature, which can lead to a feeling of inner peace, so often lacking in our increasingly technological world.

It is not only the products of bees which do us so much good: simply working with them can be enough.

BASIC FACTS ABOUT BEES

The Structure of the Bee Colony

Zoologically speaking, the bee belongs to the class of insects with a well-defined body structure, comprising head, thorax and abdomen. This kind of structure is found many times in the bee's world.

A bee colony, for instance, is made up of three different categories of bee: the queen, the worker bees who form the greater part of the colony, and the drones. The development of the bee into an adult, and its adult life, can also be divided into segments, as indeed can the housing in which the bee lives.

The bee's home is not just a simple hollow shell, but consists of several combs each containing many cells. Not all the cells are identical; there are three types: worker cells, drone cells and queen cells.

The combs themselves generally have three distinct areas: the brood area in the middle, then a circle of pollen and, on the outside, furthest from the entrance, the honey cells.

Bees also seem to prefer to live in a structured landscape with an alternating pattern of trees, hedges and fields.

Pollen, which is essential for the bee's development, is also highly structured. A single apple blossom contains about 100,000 pollen spores and a hazel nut blossom as many as 4 million. It is not surprising that pollen substitutes like soya, yeast and milk powder are no real replacement for natural pollen.

In summer the colony consists of the queen, 20,000-40,000 worker bees and 300-3,000 drones. The winter colony contains the queen, about 10,000 worker bees and sometimes, in rare cases, one or two remaining drones.

To thrive, bees need warmth and sun, nectar or honeydew, water and,

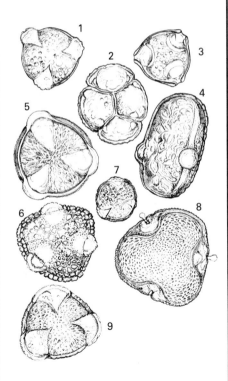

Pollen grains, greatly enlarged: 1 white clover, 2 lime, 3 hazel, 4 broad bean, 5 sycamore, 6 holly, 7 willow, 8 lime, 9 raspberry

of course, pollen. The temperature needs to be 8°C (46°F) or above for the bees to be able to fly out to collect water for the brood. For foraging, the bees require an outside temperature of at least 12°C (54°F).

One special feature in the bee world is the division of labour. The queen simply lays the eggs and leaves the worker bees to take over the care of the brood until the new bees emerge.

Bees generate warmth by increasing their rate of metabolism in the muscles of the thorax. They can very rapidly increase their body temperature by several degrees within a few minutes, something they often do before leaving the hive. Single bees do, however, rapidly lose body heat outside in the cold air and if a sudden blast of cold wind hits them once outside, they become moribund at about 7°C (45°F).

In the brood nest, the colony's 'sanctuary', where the bees live together in their thousands, a temperature of 35°C (95°F) is maintained, whatever the conditions outside. At the slightest drop in temperature they increase their metabolism and warm up their bodies. They can be up to 10 degrees warmer than their surroundings, acting as miniature 'ovens' giving off the heat they generate.

On hot days the bees space themselves out on the combs. If the heat goes on increasing, the worker bees spread a film of water over the combs and fan their wings to make it evaporate. Thus, the 'ovens' turn themselves into 'air conditioners'.

Bees share this refined art of temperature control with humans, mammals and birds, and are in this a special case because insects, being cold-blooded creatures, normally modify their activities according to the ambient temperature.

The bee colony develops as the year progresses and the sun rises higher in the sky. At the end of January, shortly after the winter solstice, the queen starts laying her eggs. In May/June when the sun is approaching its zenith, activity in the colony increases, reaching its climax around midsummer's day. By the beginning of August about 90 per cent of the brood has been laid and 90 per cent of the pollen brought in.

Egg, Larva and Pupa

Similar to butterflies, all three types of bee develop in the same four stages. The simple egg becomes a wormlike creature called the larva or grub; this in turn develops into a pupa, which, after a given time, becomes the final insect.

The queen uses a sticky secretion to glue each egg to the base of a cleaned cell. The egg is 1.3-1.8 mm in size and stands vertically. Within the next three days it bends over on its side and turns into a larva. The larva eats, grows and sheds its skin four times. The round grub becomes elongated with its head lying towards the top and its rear at the bottom of the cell.

At this stage the cell is capped with old porous wax. While stretching, the larva spins a cocoon for itself and, virtually filling the whole cell, becomes a pupa. The pupa needs to shed its skin twice more. In peace and warmth the finely structured insect develops. First the eyes become more

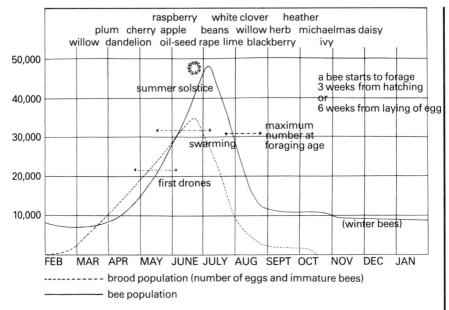

raspberry   white clover   heather
plum cherry apple   beans willow herb michaelmas daisy
willow dandelion oil-seed rape lime blackberry   ivy

summer solstice

a bee starts to forage
3 weeks from hatching
or
6 weeks from laying of egg

maximum number at foraging age

swarming

first drones

(winter bees)

FEB MAR APR MAY JUNE JULY AUG SEPT OCT NOV DEC JAN

- - - - - - - - brood population (number of eggs and immature bees)

———— bee population

Activities of a colony of bees in the course of a year

defined, then the thorax turns brown, followed by the abdomen. Only after the pupa has shed its skin for the last time do the wings develop.

Once the pupa has metamorphosed into the imago, as the adult insect is called, it is ready to emerge. The worker bee bites through the cell cap, starting in the middle, while the drones and queen attack it at the edge. The cap comes off as a round wax cover which the bees dispose of, and the pupal envelope remains behind in the cell.

The feeding of the brood starts at the larval stage. The older nursing bees have glands in their heads which produce the **brood food** or **bee milk** to feed the larvae of all three types of bee. During the first three days the larvae are given plenty of brood food. After the third day the food for the worker and drone larvae is supplemented with honey and pollen. The queen larvae receive brood food (**royal jelly**) until they emerge. It is different in both quantity and composition from the brood food given to the worker bees.

It is interesting to note that it is not until a larva has reached the elongated stage that it can defecate on the floor of its cell.

The Three Castes of Bee

The Queen

A colony is only complete when it has a queen. The queen is the focus of all the other bees. An entourage of about

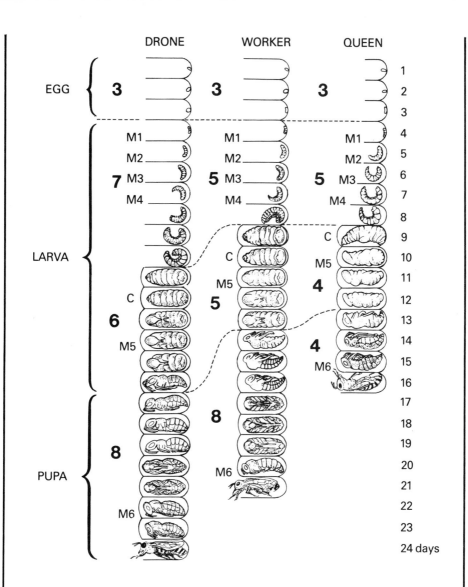

DRONE   WORKER   QUEEN

EGG
3   3   3

1
2
3

LARVA

M1 M2 7 M3 M4   M1 M2 5 M3 M4   M1 M2 5 M3 M4

C 6 M5   C M5 5   C M5 4 4 M6

4
5
6
7
8
9
10
11
12
13
14
15
16

PUPA

8 M6   8 M6   17
18
19
20
21
22
23
24 days

M1–M6=moults
C=cocoon spun

Stages in the growth of the different castes: queens mature fastest, then workers, while drones develop more slowly

Emergency queen cells raised on the face of a comb from enlarged worker cells

twelve bees, constantly changing, accompanies her at all times. They lick her, touch her with their feelers, encircle her during egg-laying and provide her with food.

The queen determines the composition and the mood of the colony. She is the only completely female member and lays the eggs from which the bees develop. In the height of the season she can lay up to 1,500 eggs per day, more than her own body weight.

All the bees have a constant and special relationship with their queen. The queen secretes a special substance, or pheromone, from her mandibular glands, which she passes on to her court and hence to all the other bees.

If the queen goes missing or dies, the colony notices her absence within the hour and starts to wail. This despairing buzzing alerts the beekeeper to the fact that the colony has lost its queen and it is always a relief when he spots her again. But how does he recognize her?

The queen's abdomen contains her ovaries and is bigger and more pointed than that of the worker bee. Her wings, therefore, are shorter than her abdomen. The queen's head is

The three castes of bee: actual (above) and in caricature (below)

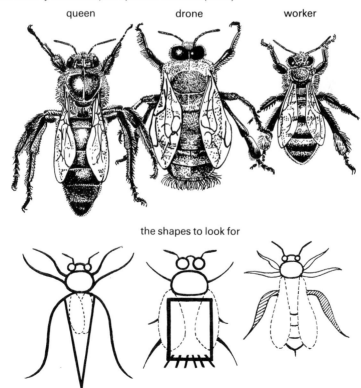

queen   drone   worker

the shapes to look for

Above: *Marking a mated queen – blue for years ending in 0 or 5, white 1 or 6, yellow 2 or 7, red 3 or 8, green 4 or 9*

Left: *Queen bee with workers attending her*

Below: *Drones driven to the hive floor ready for eviction*

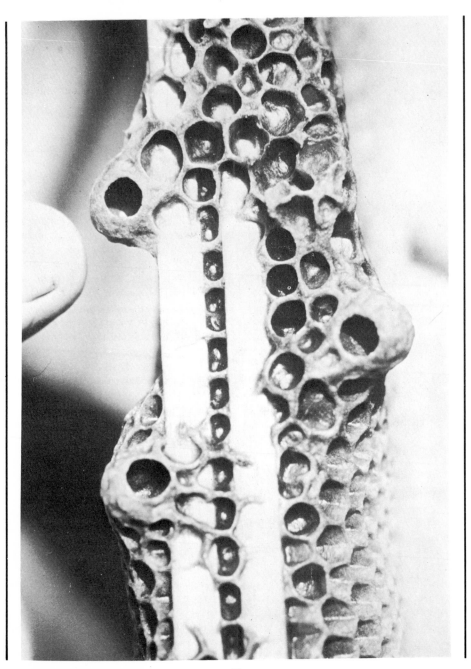

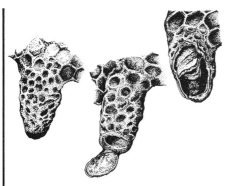

Queen cells: (left) still sealed, (middle) vacated with 'trap door' still attached, (right) bitten through

rounder than that of the worker bee, which is triangular in shape. She has no pollen baskets on her hind legs and no wax or scent glands. She seldom makes use of her sting.

She also differs from the other bees in terms of her birth and development. While workers and drones are raised in hexagonal cells, the queen has a round cell made specially by the other bees. Under normal circumstances bees always build empty queen cups on the bottom of combs. If the queen then lays in them when swarming time comes, the bees will elongate the cups into an acorn shape.

While the pupae of the workers and drones all lie horizontally in their cells, the queen pupa lies with her head at right-angles to the earth's surface.

The three categories of bee take varying amounts of time to develop. Although the queen is the largest of all, she needs the shortest time, just sixteen days. Worker bees take twenty-one days to develop, and

Queen cups on the bottom edge of a comb

drones twenty-four days, sometimes slightly longer or shorter, depending on the temperature.

There are three types of queen, depending on the conditions under which they are raised:

1 Emergency queen When the old queen dies suddenly, for instance when the colony is inspected carelessly, the bees quickly discover the calamity and select a new queen from among the eggs or new worker bee larvae. They convert the chosen cell into a round shape and look after the larva as they would the queen. Often they will construct two, three or many more queen cells at the same time.

The beekeeper can make use of this fact if he wants to find out whether a colony has a queen. He hangs a comb with eggs from another colony in the hive which appears to be queenless, as it has no open brood or eggs, and waits for one or two days. If he finds a queen cell has been started, the colony has no queen.

If he has no other queen available in these circumstances, he can leave the queen to develop. However, he must ascertain that the comb in which the queen is developing has come from a healthy colony.

Often queens in this situation are a stop-gap solution and of poor quality.

2 The swarm queen Swarming is a method used by the colony to reproduce itself. When a colony is in swarming mood, it will often produce up to twenty queen cells. Generally these are located sideways-on to the combs or on the underneath of the frame. As soon as the first cell is

Far left: *Worker pupae with cappings removed*

Left: *Foraging bee passing nectar to a house bee*

Below: *Workers gathering water to dilute honey in spring*

capped, the old queen leaves the hive together with part of the colony, forming a prime swarm.

The new queen emerges after about eight days and makes piping sounds, to which the other queens, still in capped cells, quack in reply. Now the new queen can swarm with a further section of bees, creating an after-swarm, or else rip the other queens out of their cells. Swarm queens can be of varying quality.

3 Supersedure queen When the queen becomes old and weak, the other bees take note and start to build one or two new queen cells. These are usually in the middle of the comb in which the old queen lays eggs. A new queen then develops while the old one still carries on.

The new queen is not regarded as a rival and is not stung to death. She can emerge from her cell unhindered, and at this stage two queens can exist in one colony. After about five days, the virgin queen is ready to mate and she flies out, weather permitting, on her mating flight.

If she returns, she will be ready to lay eggs after about two days; in this case there may be two brood nests in the same colony.

Sooner or later the old queen will disappear and the young queen will form the heart of the colony. In rare cases two queens have been known to live together in one colony for a long period of time, but always with their own brood area.

This form of requeening is called efficient superseding. The beekeeper will be pleased because brood activity can carry on without interruption and the colony remains strong and productive. Generally these queens are the best. The whole colony wants them and they are particularly well looked after.

What the beekeeper must remember is that the newly emerged queen cannot begin laying eggs immediately; she needs five to ten days to mature and be ready for mating. Providing the temperature is about 15°C (59°F), she will start making short orientation flights of five to ten minutes.

Only if the weather is warm and sunny will she embark on her mating flight which usually takes place between midday and 5 p.m. and lasts about twenty to sixty minutes. She flies to the areas where the drones have congregated and mates with several of them while in flight.

She receives roughly 4 to 5 million sperm cells deposited in her spermatheca. The sperm remain fertile for up to five years.

The newly mated queen can commence laying eggs after two to three days. Before laying an egg in a clean cell, she feels the size and shape of the cell with her front legs. In the larger drone cells she lays an unfertilized egg, in the smaller worker and queen cells, fertilized eggs. Fertilization occurs during egg-laying when the spermatheca releases sperm which penetrate the egg.

If the weather is overcast and the wind strong, the queen cannot mate. If the weather still prevents the queen from undertaking her mating flight five to six weeks after her emergence from the cell, she loses her fertility

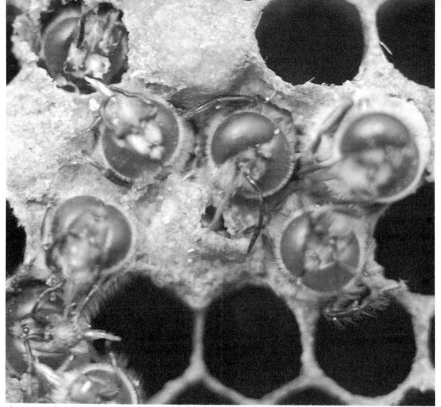

An unmated queen may still lay eggs, but they will all hatch into drones. If she lays them in worker cells, the result is shown here – drones, with their distinctly large eyes, emerging from the worker cells which are clearly a tight fit

and can lay only unfertilized eggs. If this happens, the colony ends up with too many drones.

Drones

A beginner watching a beekeeper at work will often call out delightedly: 'There's the queen!' The beekeeper will probably grin and tell him that it's 'only' a drone. The layman has, however, spotted quite correctly that it is different from the worker bees.

The drone, like the queen, is larger than the workers. He is distinguishable by his large compound eyes. He has a dumpy-shaped abdomen with his wings on top, giving him a rather benevolent, if not somewhat ponderous appearance. He can emit only a low humming noise which often frightens the uninitiated. He is quite harmless, however, having neither venom nor sting, and can be handled and stroked without fear.

The drones are the male insects whose function it is to mate with the queen. It is ironic that they are needed as fathers, as they have no father of their own, having been born from unfertilized eggs. You will generally find them in the hive between April and September.

As the drones are unable to collect

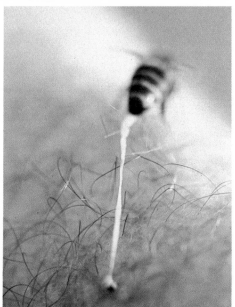

Above: *Hybrid bees. Yellow and brown bees seen together on a comb*

Left: *The sting embedded in the skin of the forearm*

Above: *Schoolchildren learning beekeeping – extracting and bottling honey*

Right: *Hives crushed by a fallen oak in a hurricane in 1987*

either pollen or nectar, and simply help themselves to the food supplies collected by others, they are often regarded as useless gluttons and lazybones who simply take their pleasure of the queen.

However, they are not entirely useless, because besides mating with the queen, they also help to regulate the temperature in the colony. When the worker bees leave the hive to collect water, nectar or pollen in the morning while it is still cool, the drones gather together on the brood combs in order to maintain their temperature at the required 35°C (95°F).

Not until the outside temperature has risen enough to suit conditions inside the hive do the drones leave the hive, and then only on sunny days between 1 and 5 p.m.

The drones also lend a hand in the honey chamber by helping to pass round the food.

Every colony has a different number of drones. If a colony finds itself short of food, the workers will throw the drone larvae out of the hive or eat them.

Drones gather in special 'drone congregation areas'. These were described for the first time by the French beekeeper Jean Probst in 1958. The drone congregation areas are in the same place every year: preferred spots in the air, with a diameter of about 50-200 m (55-219 yd), where up to a thousand drones will wait, at some distance, for the arrival of the queen, whom they recognize by her smell.

Mating takes place in sunlight in free flight. The drones often follow the queen to a height of over 15 m (49 ft), and those who succeed in mating will be the best fliers. In mating, however, the drone's life has fulfilled its usefulness and he dies at the climax of his existence. The drone, therefore, gives not only his sperm, but also his life for the continuance of the species.

Drones who do not meet their end during mating have an equally tragic fate. After blossom time, when the seeds and fruit ripen, the bees start to prepare for the coming winter. The drones, who up to now have been allowed to come and go freely and have even been fed, are kept away from the combs and the food supplies by the workers. They are bitten, pinched, stung and thrown out of the colony.

The massacre of the drones comes into full swing in August. At this time of the year you can see drones outside the hives, flying from entrance to entrance and begging, unsuccessfully, to be let in. You will often see the ground in front of hives scattered with the bodies of these starved drones, providing welcome titbits for the mice.

For the beekeeper the dead drones are a good sign. They tell him that the colony has a queen and is happy with her.

Worker Bees

In both summer and winter there are far more worker bees than others in the colony. These develop from a fertilized egg and are female in gender, but their reproductive organs are undeveloped. They devote themselves selflessly to their family; in the

Lifespan of a Worker Bee in Days

Period of service as house bee	1 2	Cleans cells and keeps brood warm	
	3 4 5	Feeds the older larvae	
	6 7 8 9 10 11	Feeds the youngest larvae	Royal jelly glands in head are particularly active
	12 13 14 15 16 17	Produces wax, builds comb and carries around food	Wax glands under abdomen are particularly active
	18 19 20 21	Guards the hive entrance	Venom glands fill up the venom sac
Period of service as field bee	22 23 24 25 26 27 28 29 30 31 32 33 34	Pollinates plants, collects pollen, nectar, propolis and water	
	35-45	Dies	

Above: *Skeps in bee boles exhibited at the National Agricultural Centre, Stoneleigh. Rape field in the background.*

Far left: *Removing a crown board over a log hive containing a colony established by a swarm*

Left: *Buying bees in nucleus boxes at a bee auction*

A drone (centre, with large eyes) surrounded by workers

altered in line with the colony's needs. When there is a good honey flow, young house bees may go out collecting, or older field bees may take up nursery duties again if the brood has to be looked after in the absence of house bees.

As soon as the little greyish-coloured worker bee emerges from her cell, her first duties in the hive begin. First she preens herself, then she cleans her own cell and other cells around her. The floor and walls of the cell are licked clean and painted with a glandular secretion, leaving them gleaming. During breaks from cleaning, the young bee warms the brood by sitting on the brood cells.

After two or three days her cleaning duties are finished and she begins to feed the brood. First of all the young nurse bee feeds the older brood with pollen and honey from the food supplies in the hive, consuming much pollen herself to meet her own needs. In so doing she builds up her brood-food glands in order to be able to supply the youngest and most delicate larvae with protein-rich brood food in the following days. One nursing bee can rear only two to three larvae.

During this nursery service, she starts taking orientation flights at midday in sunny weather. The young bee points her head to the hive entrance and dances in front of the hive, memorizing the whole of the surrounding area in five minutes.

During the second half of her time within the hive the worker bee starts building. The bees secrete fine,

course of their lives carrying out a very wide variety of duties, as the table illustrates.

The life of the worker in summer is characterized by a cycle of three × twenty-one days. The worker takes twenty-one days to develop in the cell (from egg to adult bee), lives twenty-one days as a house bee, and a further twenty-one days, approximately, as a field bee.

Tasks are allocated according to age, but the division of labour may be

Worker bees on a honeycomb with stored honey and pollen

transparent flakes from their wax glands which do not begin functioning properly until the brood-food glands start to deteriorate.

Other duties of the house bee include receiving pollen and nectar from returning field bees, distributing it among the cells, tamping down the loose pollen spores in the cells with her head and mandibles, carrying honey around and capping the mature honey stocks. She will also drag dirt and debris, such as wax remains, dead bees or wax moths, out of the hive.

Before becoming field bees after twenty-one days, the house bees take their turn standing guard at the hive entrance at about eighteen days old. Every incoming bee is checked and identified by its odour to see whether it belongs to the colony or is a robber bee. Foreign bees are only allowed to pass if they bribe their way in with nectar. Wasps and hornets intent on robbing the hive are bravely driven off by the guards.

The final duty for the diligent bee is foraging outside the hive. The maximum foraging radius of a colony is 4-5 km (2-3 miles), the average being 2 km (1 mile). Scout bees go out in search of an abundant source of food. By performing certain dances on the combs, they tell their sisters where, how far and how plentiful the food supply is.

The bee uses its proboscis to suck the nectar out of the flower's calyx into its honey stomach. The pollen it brushes into the baskets on its hind legs, which are also used to transport the propolis. It carries water to the hive in its honey stomach.

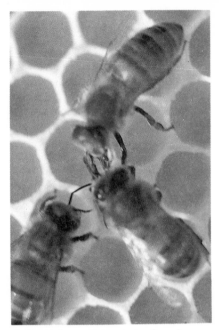

A foraging bee passing nectar to a house bee

Summer bees have a short life, lasting only five to six weeks. Winter bees can reserve all their energy for building themselves up, rather than feeding the brood, and so can live for four to nine months. A queen can live up to a maximum of four to five years, but nowadays she is generally replaced every two to three years to ensure that the colony remains fully productive.

However, it is the worker bees, as much as the queen, who determine the development and productivity of the colony. They decide, by the amount of food they give the queen, and the number of cells they clean, how many eggs the queen can lay. They also influence whether more worker bees, drones or even new queens are to be produced.

The Anatomy of the Bee

General Structure

The outward form of the bee reflects the harmonious structure within the bee world. The body is divided into head, thorax and abdomen, all in perfect proportion to each other.

The skeleton Insects do not have their skeletons on the inside like humans and mammals; they have an external framework or exoskeleton, made of a substance called chitin, which is covered in hair. The coat consists of a short, dense underpile and a top-coat of hairs which stand more or less upright. The bee uses many of these hairs to feel with. The coat varies according to the location of the bee and its state of health.

The head is a flat, thick-walled capsule which is triangular in shape, except for drones who have a rounder head. It carries the sensory organs: eyes, feelers, mouth parts and vital glands and is also the central part of the nervous system: a simple brain, which is a ganglion, being situated above the oesophagus.

The thorax anchors the two pairs of wings and three pairs of legs and thus controls the bee's movement. The muscles and respiratory system are also located here.

The structure of a honeybee

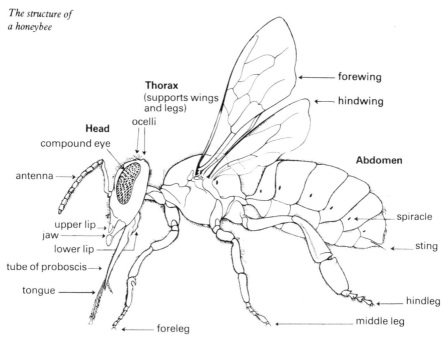

Thorax (supports wings and legs)

Head
ocelli
compound eye
antenna
upper lip
jaw
lower lip
tube of proboscis
tongue
foreleg

forewing
hindwing

Abdomen

spiracle
sting

hindleg
middle leg

Close-up of a drone

The abdomen is divided into nine segments. Each segment has an upper (dorsal) and a lower (ventral) plate (also known as tergites and sternites) which overlap like roof tiles and are connected by a finely folded membrane, allowing the abdomen to expand and contract. The abdomen can move freely in any direction as it is connected to the thorax by a small waist or 'pedicel'. The locomotory muscles for the abdomen are located under the rear thoracic segment. Inside the abdomen are found the honey sac, the proventriculus, the midgut, hindgut and rectum, the malpighian tubules, sting, venom sac, heart, wax and scent glands.

The Head

The eyes Like all insects, the bee has two compound eyes, made up of a number of single eyes, and three simple eyes or ocelli. In the worker bee the ocelli are located on the top of the head, but in the queen and drones they lie further forward. The three ocelli react to different intensities of brightness. The bees use the ocelli in the poor light conditions of the hive and also to help control pitch and roll when in flight.

The compound eyes are used for the actual process of seeing. That is not to say, however, that a bee sees as humans do: the bee can use its compound eye to perceive other parts of the light spectrum. The colour red it sees as black or dark grey; orange, yellow and green are similar. Within the ultraviolet range, which is invisible to us, it is able to perceive many nuances of colour.

The beekeeper can help bees to distinguish their own hives by painting the fronts black or red, blue, yellow and white.

The antennae The two antennae are located on the front of the forehead. In the worker bee and queen they are divided into eleven segments, while the drone has twelve. The bee uses its antennae as a refined type of 'nose' to smell and feel. It is also thought that the bee uses them to sense warmth and humidity. The individual segments of each antenna are subdivided into tiny sensory areas consisting of fine hairs. The worker bee has 3,000 of these receptors, the drone as many as 15,000. Each of these fine hairs is connected by a nerve to the brain.

The sense of smell is vital to the bee both in the dark inside the hive and outside in the daylight. The bee uses its sense of smell to identify its colony, water, plants and possibly even the beekeeper himself.

As far as we know, the bee uses its eyes to identify objects at a distance and its sense of smell for nearer objects.

The mouthparts The front jaws or mandibles are pincer-shaped and point outwards. After the tongue they are the most important part of the mouth. The bee uses them to open the cell cap when it is time to emerge from its cell. They are also indispensable for collecting pollen and honey, feeding larvae, manipulating wax and propolis, gripping other bees and carrying rubbish.

The mandibles are strong enough

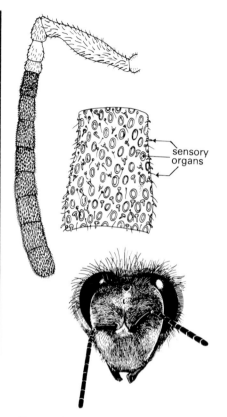

sensory organs

Close-up of a bee's antenna: the head of a worker bee, with antenna enlarged, and a segment still further enlarged

to bite through paper, soft fibreboard or polystyrene, but not through fruit skins. Smooth-surfaced fruits need to be damaged before the bee can suck at them, whereas wasps can use their sharper mouthparts to attack even undamaged fruit.

The mandibles are different for the three categories of bee. Strangely enough it is the queen, not the worker bee, who has the strongest and sharpest mandibles. The drones have the weakest of the three.

37

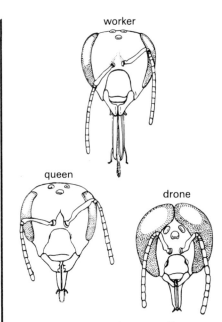

worker

queen

drone

The 'face' of each of the three castes of bee

The proboscis The proboscis is another important organ in the mouth. When the bee is at rest, it is retracted so that it cannot be seen. When lapping or licking, the bee unfolds it and dips it in the liquid.

The proboscis consists of five parts, four of which are connected to an airtight sucking tube. In the middle of this tube is the tongue which projects beyond the tube. It is covered in thick hair and can be moved in all directions. At the end of the tongue is a delicate spoonlike lobe called the labellum which the bee uses to gather up even the minutest quantities of nectar.

By retracting its tongue, the bee introduces the nectar into the sucking tube and then employs a pumping action to suck it up into its mouth

cavity, through the pharynx, along the oesophagus and into the crop or honey sac.

The length of the proboscis, which varies from 5.6-7.1 mm (⅕-¼ in), determines the type of plant the bee visits. Only Caucasian bees and some Carniolan strains which have slightly longer tongues can suck the nectar from red clover florets with their long corolla tubes.

The tongue also has another fine groove on the inside, along which the bee can pass saliva to mix with honey and hard sugar paste. This enables the bee to make the viscous honey (with its 80 per cent sugar content) soft enough to digest.

The bee tastes through its mouth parts, although its sense of taste is different from a human's. Things that taste very bitter to us do not worry bees, and they are less sensitive to sugar. To us even a 2 per cent sugar solution tastes sweet, whereas the bee is interested only in solutions of 4 or more per cent, preferring higher concentrations (the sugar content of nectar varies from 5-70 per cent).

The bee's proboscis, seen from the rear with its component parts spread out

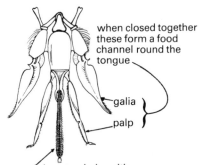

when closed together these form a food channel round the tongue

galia

palp

tongue in extended position

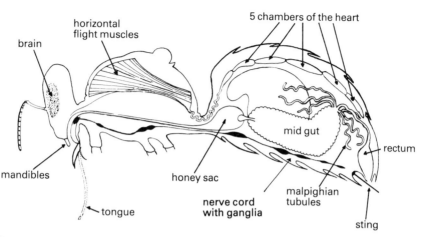

Labels on the diagram:
brain
horizontal flight muscles
5 chambers of the heart
mandibles
mid gut
rectum
honey sac
tongue
nerve cord with ganglia
malpighian tubules
sting

The main internal parts of a worker bee

The salivary gland consists of five interconnected branches. One branch lies behind the brain, the other four between the flight muscles in the thorax. The discharge from all five flows into a small pocket above the tongue. From there saliva flows down the groove in the tongue to moisten the honey.

The saliva is thought to play a part in the digestion of fat from pollen and to supply additional materials to nourish the brood. The salivary gland also produces the substance used by the larva to spin its cocoon.

A double function is also performed by the **brood-food glands** (the pharyngeal and hypopharyngeal glands) which run into the lower part of the pharynx. When fully developed, these glands fill most of the bee's head between the brain and forehead wall. Only vestiges of these glands are found in the queen bee, and the drone has none at all.

The glands comprise two long tubes lined with several hundred tiny glandular sacs which are fully developed at the time the young bee is looking after the brood. The nursing bees use these glands to make the brood food or royal jelly for the youngest larvae and the queen. The older bees secrete an enzyme with the food which separates nectar sugar into fructose and glucose.

The Thorax

The head is connected to the thorax by a narrow flexible neck.

The wings are attached to each side of the middle and rear segments of the thorax. They are constructed of membrane and have a framework of small veins. The type of veining serves as a racial characteristic for the breeder. The forewings are larger than the hindwings. In flight the wings are linked together by a series

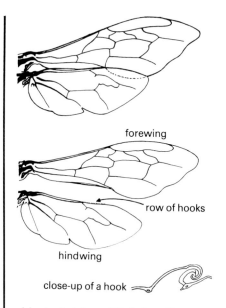

forewing

row of hooks

hindwing

close-up of a hook

A bee has both fore and hind wings. The wings are hooked together in flight (top) but separate when at rest (below)

of hooks; wings linked together in this way can beat 250 times per second and, in wind-free conditions, achieve a speed of about 29 km (18 miles) per hour.

In flight the wing tip describes a figure of eight, with the leading edge of the wing providing directional control. At rest the wings are unhooked and lie on the bee's back, folded lightly together.

All three pairs of legs are highly articulated and each pair is different. Each leg comprises six individual segments which are connected to each other by flexible joints. This allows the legs to move in various directions and at various angles.

Drone with wings unfolded, showing both pairs

The forelegs have an antenna cleaner which is a deep, round notch fringed with stiff hairs; directly above this notch is a spine which acts as a support. After each collection of pollen the bee draws its feelers through the antenna cleaners which collect pollen spores and dust particles from them. The bee does this at regular intervals, even when the feelers, with their many sensory organs, are not dirty.

The middle legs have the least specialized function. They help in walking and brushing off and passing on pollen from the bee's body. The bee can also use them to press the pollen and propolis load firmly onto its hindlegs from the outside.

The hindlegs are the most highly developed and specialized. They carry special collecting equipment: brush, comb, basket and pollen press. The brush on the basitarsus (basal joint of the foot) consists of about ten rows of stiff, backward-sloping bristles. This brush collects the pollen combed off by the front and middle legs and also the pollen brushed off the abdomen by the back leg itself.

If you watch a foraging bee lift off from a flower, you will see it quickly moving its two hindlegs against each other over the flower. In doing this the bee moves the powerful comb at the outer edge of the tibia of one leg through the brush on the other leg, collecting the pollen in the comb. The basitarsus, which has a spoon-like hollow with teeth pointing upwards above the brush, moves against the tibia.

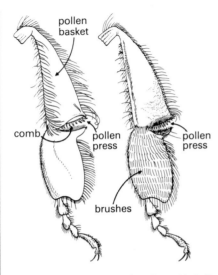

A worker bee's hindleg, seen from the outside (left) and inside (right), showing the pollen press

The pollen press squeezes the pollen collected in the comb through the joint, so that it moves from the inside of the thigh to the outside, where the pollen basket is located. This holds the growing pollen mass in such a way that it remains fixed during flight but can easily be removed once the bee is back in the hive.

The Abdomen

The digestive organs However complicated the outside of a bee may seem, its internal organization is simple. The alimentary canal is a long, narrow tube running from the head and thorax to the front part of the abdomen and from there into the honey sac. The honey sac is connected via the proventriculus to the stomach, which is also the midgut.

The midgut leads into the hindgut – a narrow tube with a sac-like rectum and rectal gland. The rectum is connected to the anus, located between the sting and the last dorsal segment.

The honey sac or crop is a delicate, transparent membranous sac, surrounded by muscle fibre, which leads into the alimentary canal. It can expand quite considerably; when packed full it is about the size of a match head and weighs between 40 and 70 mg.

The beekeeper can tell whether the bee's honey sac is full by looking at the bee's abdomen as it lands on the alighting board: if full it will be dipped low.

If field bees take a rest before marching back into the hive, that is a sure sign they have had a fruitful journey. To find out whether a bee is bringing nectar or water, simply put

The bee's honey sac is located at the entrance to the gut. It is shown here in section (above) and in a sketch overview (right)

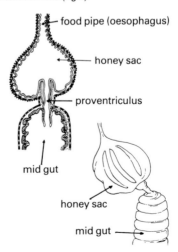

the bee on a fingernail, press lightly on the abdomen and taste the drops that come from the bee's mouth.

The bee uses its honey sac to transport not just nectar, honeydew and water, but also 'fuel' (honey) for its flight.

The proventriculus connects the honey sac to the stomach and digestive system. It is a double-spouted tube, one end closed by a four-lobed, one-way valve leading out of the honey sac, the other end feeding into the stomach. Its function is to allow the food needed by the bee for its own nourishment to pass into the digestive system, but to prevent any returning to the honey sac, thus ensuring that the honey remains free of any digestive waste.

The midgut is a tube which acts as both stomach and intestine and is used for the digestion of food.

The intestinal wall secretes digestive juices which are continually mixed with the partly digested food by the action of the sphincter muscles.

The malpighian tubules are thin tubes, between the midgut and hindgut, which coil around the body cavity. They stretch around the whole of the abdominal chamber where they are bathed by the freely circulating blood fluid. They withdraw waste products and salts from the blood and discharge these into the hindgut and thence into the rectum.

The hindgut continues to absorb nourishment from the partly digested

food, or chyme, passing it on to the blood.

The rectum can expand a great deal to absorb large amounts of undigested matter, a fact which is particularly important in winter. The rectum may fill the whole of the space in the abdomen, pressing all the organs close up together. A healthy bee will only empty its rectum outside in the fresh air.

The sting and venom sac The sting is an intricately jointed piece of equipment, consisting of about twenty parts.

It has two pointed lancets, the edges of which are lined with re-curved teeth or barbs. At rest the sting is kept in a retracted position in its sheath in the abdomen. Located close by is the poison sac which stores

The sting mechanism revealed

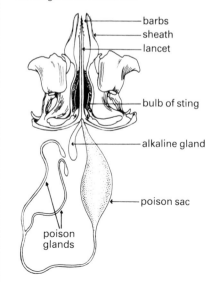

- barbs
- sheath
- lancet
- bulb of sting
- alkaline gland
- poison sac
- poison glands

the venom, and two venom glands.

When the bee stings, its abdomen bends rapidly downwards, driving the sting out of its sheath and penetrating the victim's skin with a jerk. The barbs on the lancets are now embedded in the skin, so that if the bee flies away in this condition, its venom sac and all the venom glands and the ganglia will be ripped from its body and the bee will later die from its wounds. The sting, however, is thus driven by reflex action further into the wound.

If you are stung, you should scrape the sting off with your thumbnail to avoid squeezing the contents of the venom sac into your skin. If a bee stings another insect, it can usually retract its sting from the insect's coating of chitin and therefore remain alive. The queen's sting is slightly longer than the worker bee's, but has fewer barbs on the lancets. She seldom uses it. Drones have no sting.

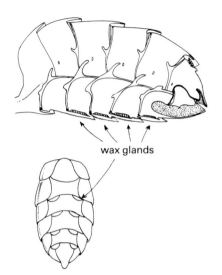

wax glands

A section view of the abdomen, showing the wax glands in which the wax scales (below) are formed

The wax glands Bees are able to manufacture their building material themselves in glands through which they secrete wax. These glands are

The main glands of the honeybee

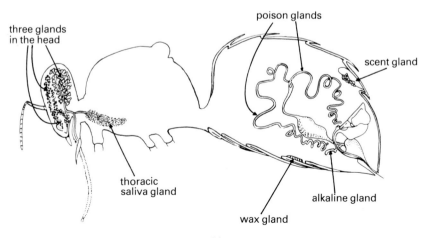

three glands
in the head

poison glands

scent gland

thoracic
saliva gland

alkaline gland

wax gland

A worker bee fanning to distribute scent from her scent gland, which is seen near the tip of her abdomen

located in pairs on the underside of the final four abdominal sternites.

They are, in fact, specialized sections of skin cells, each containing between 10,000 and 20,000 glandular cells which become mature and fully developed once the bee is old enough to start building work. The wax, which at this stage is liquid, is secreted through fine pores in the 'wax mirrors' covering the glands; between the wax mirrors and the plates it hardens into transparent scales.

The scales are loosened and discharged from between the segments to be gathered up by the hindlegs, passed forward to the mandibles and masticated to make them malleable. Only now is the wax ready for comb building.

The scent gland or Nasonov gland, is located between the penultimate and last abdominal tergites. Only in worker bees is this gland fully developed. The bee raises its abdomen and, in a process of communicative fanning, secretes a delicate odour smelling of lemon balm. This odour helps young bees, on orientation flights, and returning field bees to find their hive again.

Heart and blood circulation The bee has an open circulation system with only one blood vessel, the heart. Located in the abdomen, the heart consists of a thin tube with five intercommunicating chambers which suck the pale-coloured blood fluid from the surrounding area and pump it from the abdominal end up to the head.

The vessel ends in the head, so that the blood flows out freely into the body cavity and filters back slowly to the abdomen. In doing so, it flows round all the internal organs and nourishes them.

Respiratory system Bees have an extensive system of coil-like breathing tubes, or tracheae, which come from large air sacs in the head,

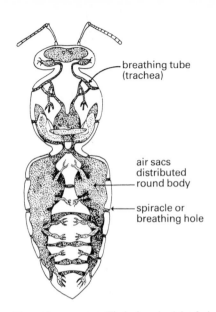

The respiratory system. The bee's main air intake is not through its mouth, but through breathing holes in its skin (external skeleton)

thorax and abdomen; some are also located in the legs. They form an intricate network which penetrates all the organs, supplying them directly with oxygen. The system, which also keeps some of the organs, e.g. the gut, buoyant in the body cavity, has openings in the side of the body, called spiracles, which are fringed with hair.

There are three pairs of spiracles on the side of the thorax and seven pairs on the abdomen. The bee uses the abdominal spiracles to breathe in and the thoracic spiracles to breathe out. Breathing is controlled by the abdomen which expands and contracts rhythmically.

The nervous system connects all the bee's sensory organs, body organs

The circulation system. A bee has a heart, but it is a simple one, consisting of five chambers that open and close for propulsion

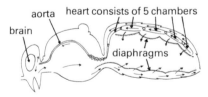

how a chamber of the
heart functions

propulsion   intake

46

and glands. The nerves run together in a central nervous system running between the brain and the abdominal nerve centres, or ganglia, to form a pattern resembling a rope ladder. Little is yet known about the precise functioning of the nervous system.

Fat body Like many other insects, the bee has a special tissue in its abdomen which is able to store fat and protein. In the first weeks of its life the young bee consumes large quantities of pollen to build up a fat-protein body. Because some eggs will be laid in the cold weather early in the following spring, the bee will need this fat body to nourish that first brood.

The reproductive organs are atrophied in the worker bee but may, under some circumstances, start to function again. The queen's ovaries are located in the abdomen. They are pear-shaped and consist of about 180 ovarioles, resembling pearl necklaces, which contain the germ cells in which the eggs develop.

The ovarioles converge to an oviduct which passes into the vagina. The round spermatheca, which is used to store the male sperm ejected into the queen during coition, is located above the vagina.

The drone's testes are located at the front of the abdomen on its side. In the sexually mature drone the sperm develop in the testes and pass through ducts to a bottle-shaped sperm sac where they are stored until mating. In the mating process the drones lose their sex organs and die shortly afterwards.

Four Bee Species

Anyone who has anything to do with animals or plants knows that for each genus there are different species, and for each species different sub-species or ecotypes. This is also the case with the bee.

Our honeybee belongs to the genus *Apis*, of which there are four species:

1 The domestic honeybee (*Apis mellifera*) was originally native to Africa, the Middle East and Europe as far as the Urals. It was taken by man to the American continent and to Australia. Today it is the most prevalent species and, under natural conditions, nests in hollow trees where it builds a home consisting of a series of combs. The European sub-species remain close to their home or hive.

2 The Indian bee (*Apis cerana*) is native to India and South East Asia and also builds its nest in hollow trees. It has a tendency, however, to swarm and leave the nest.

3 The giant honeybee (*Apis dorsata*),

4 The dwarf honeybee (*Apis florea*) are both native to India. They build only free-hanging combs. In bad weather they, too, have a tendency to abscond.

All four species are social. Bees belonging to species that build colonies can be found as far north as Siberia, although their whole life is geared to the sun, on whose warmth they depend.

47

Sub-species of the Honeybee

The **black** bee from north-west Europe *(Apis mellifera mellifera)* is a relatively large bee with a dark shiny appearance and rather sparse hair on its abdomen. It is closely related to the North African bee. It was originally native to Spain but has now spread to western, northern and central Europe and, via northern Russia and Siberia, to the Pacific. In Germany it has been largely superseded by the Carniolan bee (see below).

In spring the black bee develops slowly but then remains in peak form for a long time in the summer. It lives in medium-sized colonies and overwinters in relatively large units. It has a strong swarming instinct, tends to scramble over the comb, and uses its sting a lot. It is known for its vitality and hardiness.

The Swiss still breed and keep a strain of black bee, and it is also kept on the Lüneberg Heath in north Germany where swarming is actively encouraged. It has been known for an overwintering colony to swarm up to six times, with even the prime swarm reswarming. To do this the bees need a very marked brood instinct which carries on through the summer, producing a large winter population.

The Italian honeybee *(Apis mellifera ligustica)* is easy for a layman to recognize because of the one to three gold bands on its abdomen. The yellow colour can vary from a light to a brownish tone. The body hair is also yellowish. It comes originally from Italy and is now prevalent in parts of the USA and in other countries with similar climatic conditions.

Its spring development is average, but it retains strong colonies throughout the summer. Adapted to the Mediterranean climate, brood activity can continue right into the winter, and the Italian bee, therefore, overwinters in strong colonies and requires a lot of food.

It does not swarm unpredictably, does not scramble over the combs and does not sting too much. Nevertheless, it does have a tendency to rob other colonies and to leave the hive. When the weather conditions in spring are severe, its development is slow. In harsh regions it often has a problem overwintering and is best suited to areas with a long honey flow and mild winters.

The Carniolan bee *(Apis mellifera carnica)* has a dark-coloured body and a thick coat with broad grey bands. It is native to the south-eastern Alps, the north Balkan region and the Danube. Its development is rapid in the spring but falls off in early summer. It overwinters in small colonies and does not require much food.

Carniolans sometimes have a strong swarming disposition, particularly non-thoroughbred strains, and adapt their brood activity to the weather conditions. They are much loved because of their gentleness and the way they stay quietly on the combs. Their early spring development makes them suited to regions with an early honey flow.

Mention must also be made of some

further sub-species of honeybee:

The Caucasian honeybee *(Apis mellifera caucasica)*, which is likewise grey, comes from the Caucasus and is widespread in the USSR and central Europe. It is very adaptable and accustomed to the extreme weather conditions which are associated with the European climate. In Germany, however, Caucasians do not winter well.

The Cyprian bee *(Apis mellifera cypria)* lives in Greece and several parts of south-east Europe and is noted for its above-average fertility.

The Buckfast bee counts more as a strain than a sub-species. In 1969, Brother Adam, from Buckfast Abbey in the south of England, produced a new hybrid by crossing various other races of bee. Buckfast bees are good at brood rearing. They develop slowly in spring but brood activity continues on into the autumn. They have little tendency to swarm, sit quietly on the combs and have a gentle disposition.

On the debit side, however, is their tendency to rob other colonies and to abscond from the hive. They are not well suited to areas where there is an early honey flow and they require higher temperatures for foraging than the Carniolans.

A commercial queen-rearing apiary where mating nuclei are housed. Because it is hard to prevent unselected drones from mating with the virgin queens on their nuptial flights, pure strains soon become hybrid

STARTING IN BEEKEEPING

In ancient times knowledge about bees was the preserve of secret cults and only certain specially appointed people were allowed to keep bees. Later on guilds, too, acquired this knowledge. Now, of course, everyone has more leisure time and it is a matter of freedom of choice whether or not to keep bees. However, if you are seriously thinking about becoming a beekeeper, first ask yourself if you have what it takes.

The first and most important prerequisite is a love of honeybees. If you are really interested and want to study nature, you are already halfway there – but only halfway. A colony of bees can be very fascinating, but that doesn't mean you should launch yourself into beekeeping without any prior knowledge and without asking the advice of an experienced apiarist.

First of all there are certain basic facts and rules to learn. If you want to get it right, you have to adapt to the laws of the bees, and not vice versa!

It is vital for a novice to understand that every bee colony has an individual identity and needs individual handling. This means one colony may react completely differently from another colony to the same thing. What is right for one particular colony in one particular year may be completely wrong the following year. That's what makes beekeeping so

Pollinating apples: hives should be placed in orchards just before buds open, and removed at petal-fall to escape spray damage

fascinating; it is full of surprises!

In order to be a good beekeeper, you need inner peace, a gift for observation and powers of empathy. Often you need to act on a sudden inspiration and will feel enormous satisfaction at getting it right. This feeling is one of the special attractions of beekeeping.

Bee stings, however, are not so attractive. Any would-be beekeeper should find out whether or not he or she is allergic to bee venom. You will know you are allergic if, when you are stung, other parts of your body also react to the sting: if, for example, you are stung in the leg and your neck swells up. But an allergy will not necessarily prevent you from keeping bees; it can be cured, under medical supervision using the bee venom itself.

Other prerequisites for the small beekeeper are: a suitable location, accommodation for the bees, a small storeroom for equipment and a knack for do-it-yourself which saves money.

Suitable Locations

Before you buy your first colony, or catch your first swarm, you should consider the best place to put the bees. You must think not only what is best for you (near to home, easily accessible, reachable by car, etc.), but also what is best for the bees.

A beekeeper learns from experience that a good location will help

A college apiary at Merrist Wood in southern England

prevent many diseases. Favourable sites are often to be found in outlying areas, but places near roads with heavy traffic, or motorways should be avoided because vehicles travelling at speed stir up the air, endangering the lives of bees in flight.

In populated areas, suitable sites are large gardens and orchards, but it is important to ensure that neighbours are not disturbed and for this reason it is a good idea to keep the hives slightly screened off. It is also advisable to have a high hedge or fence between the bees and neighbouring properties, because the bees will then be forced to fly upwards to clear the obstacle, avoiding neighbours and passersby.

Not every local authority lays down regulations about beekeeping, but it is always advisable to keep a certain minimum distance from your neighbours' property. It is well worth checking with your local authority.

It is not a good idea to keep bees in a small garden attached to terraced houses or high-rise buildings, as people may be scared of them.

In gardens of 400-500 sq m (480-600 sq yd), with trees and bushes, there should be room for three to four colonies. It is a definite advantage to site the bees close to your house; you can then keep a better watch on them and you will not have so far to carry your equipment.

Ideal sites for hives are old quarries, wasteland or remote corners of woods where there are generally few problems with other property owners, though one should have

Apiary in winter

regard to the likelihood of vandalism in semi-public places. Local authorities and forestry commissions often lease land at little to no cost.

The hives should preferably face south east, protected from the midday sun. Bees hate damp and like warmth, but too much heat can lead to premature brood rearing and swarming. The leeward side of a dry slope is a good spot, as it should protect the hive from strong winds and fog.

Bees find it very difficult to take off and land again with their loads if there is too much wind. You should also make sure that the hives are not too far from the bees' source of food: if they are surrounded by flowers and woods, your combs will soon fill with honey.

How to House your Bees

The next question you are faced with is what type of housing to use for the bees. This is important because everything else depends on it. In the past, bees were kept in many different ways: in earthenware cylinders, or in hives made of woven rushes or straw, called 'skeps'; or in logs or hollow trees.

The invention of the movable frame hive, in the nineteenth century, revolutionized beekeeping and made it easier to study the happenings inside a bee colony. It meant that the beekeeper could turn the combs round, change them over, insert foundation frames, and so on.

A queen excluder to separate the brood chamber from the honey supers

mentioned here. In Germany there are two main types of hive, the rear opening and the top opening hive. Most of the rest of the world uses top opening hives exclusively. In Britain the main division is between single-walled and double-walled hives, and also between top bee space and bottom bee space hives. Skep hives are now obsolete for practical purposes. These were accessible only from the bottom.

Rear-opening Hives

Rear-opening hives are noticeably declining in use in Germany. They are complicated by having a door at the back through which frames are removed or inserted with tongs. Frames must be hung on a special rest during inspections and it is not surprising that this style of hive is not found in the UK or the USA or other regions where beekeeping is a major industry.

Even the variant of the rear-opening hive which has the frames arranged end-on to the rear door, or the one which has rails so that the frames can be pulled out and examined from above, have not been taken up in other countries.

These hives have advantages but they require to be kept in a bee-house. Also they are time-consuming to use, complicated to make and expensive to buy.

Top-opening Hives

The majority of the world's hives consist of piles of boxes standing on a floor with an entrance at the bottom.

Soon there were disputes about the best arrangement for the frames: whether it was better to arrange them parallel to the front of the hive for warmth, or end on to it for cooling. The straw skeps used neither of these arrangements, but positioned the combs diagonally to the bee entrance.

A further long-standing argument in beekeeping was the question of the best type of hive. Some apiarists even maintained that every region should have its own shape and size of hive and, considering the variety of hives they invented, were as individual as their bee colonies. Only a few of these hives which are still in use can be

Each box is without a top or bottom, but has a ledge on two opposite sides on which frames can be hung in parallel rows.

The top box is covered with a crown board and a bee-tight roof. Normally the bottom box is deeper and kept exclusively for the brood nest. Sometimes the queen is allowed up to lay in a second brood chamber at the peak of the brood-rearing season. This can be a shallow or deep box, according to the strength of the stock.

When a queen excluder is laid over the brood box any 'supers' or boxes of frames placed above it are used for honey storage only. These boxes usually carry shallow frames, spaced wider than those in the brood box to produce thick combs which cannot

therefore be exchanged with frames in the deeper brood chamber.

It is possible in a tall hive consisting of three to five boxes to create bee-tight subdivisions with their own entrances to accommodate young colonies, so saving space and heat.

The Langstroth Hive

The idea of 'honey-supering' or placing one container on top of another which contained the brood nest was commonplace, even when the latter was just a straw skep. In the mid-nineteenth century an American, Pastor Langstroth, discovered the bee space and built the original top-opening brood-box with moveable frames.

Two brood boxes with frames 'cold way' (left) and 'warm way' (right)

Round skeps or round supers were no longer suitable to take rectangular frames, and the familiar modern pile of boxes came into common use.

The Langstroth hive and variants of different sizes spread throughout the world, finding more and more enthusiasts. Commercial beekeeping would now be unthinkable without this form of hive, with its interchangeable parts, since it is a very time-saving method of keeping bees and the boxes are simple to make.

Hives used in Britain

In Britain the most common hive is now the Modified National Hive which resembles the Langstroth except that the ledge on which the frames hang is more complicated than the simple rebate of the American hive.

The National can be placed on its stand (being 460 mm/18⅛ in square) either with the frames running fore and aft to the entrance (cold way) or parallel to it (warm way). The wider ledge allows for long-lugged frames and the hive has 'bottom bee space'. This means that the top of a layer of frames is flush with the four sides of the box they hang in.

When another box of frames is put on top of it the bees sitting on the top bars below are not crushed because there is a 7 mm (¼ in) gap or 'bee space' under the frames above, which are not flush with the bottom of that box.

Dadant, Commercial, Smith, Langstroth and other single-walled, top-opening hives are all now familiar to British beekeepers, and have their enthusiastic supporters. All except the Commercial hive have top bee space.

The WBC Hive

Non-beekeepers in Britain would probably not recognize a National as a beehive. The traditional image of a 'proper' hive is the white-painted, pagoda-shaped WBC double-walled hive which looks ornamental in a garden setting for the delight of the hobbyist.

This hive takes the same frames, interchangeable with those in the National. However, the boxes take only ten frames and are constructed of lighter timber. Protection from the elements is provided by telescopic

The parts of a top-opening hive

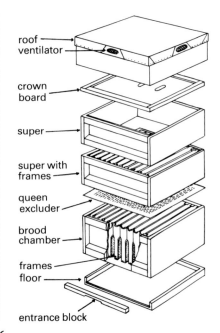

roof
ventilator

crown board

super

super with frames

queen excluder

brood chamber

frames
floor

entrance block

56

Beehives with their brood frames. From left to right, top: Langstroth, WBC; bottom: Commercial, National, Dadant

'lifts' of stouter timber which provide an outer shell.

The WBC is expensive when new (though cheap second-hand), time-consuming to open and close, difficult to keep bee-tight and most difficult to move from place to place. No bee farmer could afford to use WBCs, but many thousands of small-scale apiarists prefer them despite the disadvantages of expense and inefficiency.

The ten small British Standard frames of the WBC brood box are not capacious enough for a strong colony in a good season. Even the eleven frames of the National are inadequate, requiring a second or relief brood box below the excluder. Bigger frames, like those in the Commercial hive or the Dadant, can accommodate all the brood in one box.

However, in winter the outer frames may become mildewed if they are unoccupied when the brood nest shrinks. They also require a strong back to lift about.

Fixtures and Fittings

Single-walled hives are best for migratory beekeeping, but they must have a good security system to prevent them from slipping, and it is best to have a ventilation screen on top to close in the bees but allow some ventilation.

At other times a stout, well-fitting cover-board or quilt under the roof is essential. If it is well insulated it

Various feeders in common use, and a carton of candy (bottom left)

prevents heat loss in winter and keeps heat out in summer. Heat loss can be prevented by placing a plastic film over the frames in the topmost super.

The hive floor should be raised on a small stand and have a space under the frames in which an insert can be placed so that it can be pulled out to check for Varroa mites.

A strip of wood or a mouse excluder with 9 mm (⅜ in) holes is needed in winter to keep out mice and the entrance may need to be reduced to a single bee-way to keep out robber bees.

There are many types of feeder for these hives. Generally speaking, contact and indirect feeders in metal or plastic containers can sit over the feed hole in a cover board, protected by an empty shallow box under the roof. Larger Miller or Rowse feeders are placed on the top, too.

Otherwise, frame feeders can go directly into brood boxes, or food can be supplied through a plastic feeding bottle which can be inserted into the hive entrance.

Migratory Hives

The German Hohenheim migratory hive is specially designed for easy movement. Special hooks or clasps are provided to secure the bottom part of each box to the top of the one below, and a lock to secure the whole pile of boxes. Each box has a 60-70 mm (2.4-2.8 in) antechamber to provide ventilation space when in transit. This space comes in handy to rest the first frame when manipulating the colony. The system uses a pan feeder above the top box.

Materials for Box Construction

The material to construct the boxes is important. Common materials in use are expanded polystyrene, external quality plywood, western red cedar and good quality pine. Plastic hives are rigid, light and warm, but condensation in the hive can lead to mould forming on outer combs in the winter.

Plastic will not accept nails or staples. Plastic boxes have a relatively short life span, and the material can be chewed by bees, mice or ants, or pecked by birds. It can also provide shelter for wax moths. Carelessness by the beekeeper can lead to damage and the material can give off noxious vapours.

Wood, on the other hand, is very versatile. Find some 19 mm (¾ in) Weymouth pine. Being light and porous it lets air in and moisture out. This helps air-conditioning and allows the colony to breathe easily. Do not paint the outside with any protective coating which is non-porous or contains an insecticide.

Weymouth pine does not warp as readily as other woods. Deal can also be used successfully for hive construction.

Rye Straw

An old building material which has recently been rediscovered is rye straw. The straw walls should be about 5 cm (2 in) thick, strengthened with wooden laths and covered in a coating of cow dung and clay. Boxes like this can be made at home at little expense.

The insulation works so well that no condensation forms and there is always a comfortable heat inside the colony, even in winter. However, woodpeckers can damage the hives and so they are normally brought into a bee-house for the winter or given some other protection.

Advantages and Disadvantages of Hives with Supers

As the hives cannot be stacked one on top of the other, they take up a lot of room, but no bee-house is needed. They also require little in the way of additional accessories. The empty boxes can be used in winter to store empty combs, but you do need a place to store the boxes. Another good point is that they are easy to make.

Supers can be heavy and difficult to lift off. As there is little insulation within the hives, food consumption in winter is quite high.

On the other hand, this type of hive takes relatively little time and effort to manipulate. Jobs like inspecting the colony, expanding or uniting colonies, catching swarms and feeding can be done quickly and easily and the colony can be expanded as required by adding more supers.

Because manipulation is comparatively easy, however, a novice beekeeper may be tempted to inspect his colonies daily or to acquire too many colonies. His whole operation may then take on such proportions that he no longer has time to give the colonies the individual attention they need.

The Frame

Langstroth had a decisive influence on the design of movable frame hives. His vital discovery was the fact that if a space of 6-10 mm (¼ in) is left between the hive wall and frame at the sides, and at the top and bottom, the bees do not try to glue up the space with propolis or to build combs. The frame, therefore, remains movable. The best distance between the frames and the sides of the hive was pitched at 7 mm (¼ in), leaving the bees with just enough room to get by.

The frame is made of four wooden laths forming a rectangle. The most popular dimensions in Germany are German Zander and German Standard, but many other sizes are used. A few are listed below, together with the main frames used throughout the world. Inches (without metric equivalents) are used for the sake of neat presentation.

The best material is 8-10 mm (approx ⅓ in) thick, 20-25 mm (approx 1 in) wide wooden laths made of knotless pine, deal, linden, alder or poplar. For greater stability, the side parts of larger frames should be made from a hardwood such as beech.

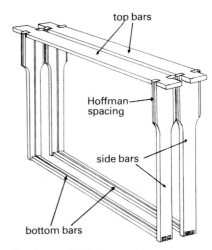

Two frames with Hoffman spacing

Drone frames are also best made from hardwood. Most people today will buy cheap machine-cut frame parts and make them up themselves.

The five laths are usually nailed together and some beekeepers glue them as well. There are push-pin

Name	External dimensions (inches; width × height)	Comb area (sq in one side)
Germany:		
Zander	16¼ × 8⅝	124
German Standard	14½ × 8¾	114
Kuntzscher	13 × 9⅞	114
Gerstung	16⅛ × 10¼	149
Other countries:		
British Standard	14 × 8½	93
Commercial	16 × 10	130
Dadant	17⅝ × 11¼	159
Langstroth	17⅝ × 9⅛	128
Victoria Deep	17½ × 9⅛	127

tools available to make the work easier and it is simple to make up a jig to ensure you get a perfect right-angle.

An important part of the frame is the spacing. The frames must be a precise distance apart: 35 mm (1⅜ in) from the middle of one comb to the middle of the next. The Hoffman system of self-spacing is in widespread use. The side pieces of the frame widen out at the top.

Foundation

Wires are stretched across the frame and a sheet of foundation embedded on them. The foundation is a sheet of beeswax on which the pattern of the base of the worker cells is embossed. The bees use this 'starter' as a guide to build up the cell walls.

The Comb

The bees build up a comb of hexagonal cells in which to raise brood and to store honey and pollen.

Some systems with very tall brood combs also use special thick combs in the honey supers. These are only half the height, but double the thickness of brood combs. As the cells are therefore twice as deep as brood cells, the queen does not lay eggs in them and no queen excluder is needed.

In the wild, bees build heart-shaped combs. The drone brood is laid in the lowest section of comb, beneath the worker brood. This is why, in top-opening hives, the drone brood is often to be found in the bottom chamber.

It is not desirable to have worker

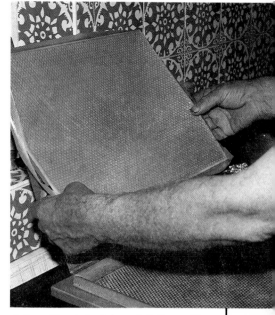

A sheet of foundation being peeled off a mould

and drone cells mixed up on a comb. To prevent this the beekeeper can hang empty, unwired frames, one placed in on the right and one placed in on the left of the brood nest, possibly with a starter strip of drone cells fixed with wax.

The bees will then use these to raise drones. The drone frames can be more rigid than normal frames, mainly because they will be drawn out first and are therefore subject to greater stress.

In July/August the beekeeper removes the drone frames and melts the wax. Some European beekeepers cut out the drone cells each time they inspect the colony, to eliminate Varroa mites which like to shelter there, and to obtain wax.

Drone cells differ from worker cells

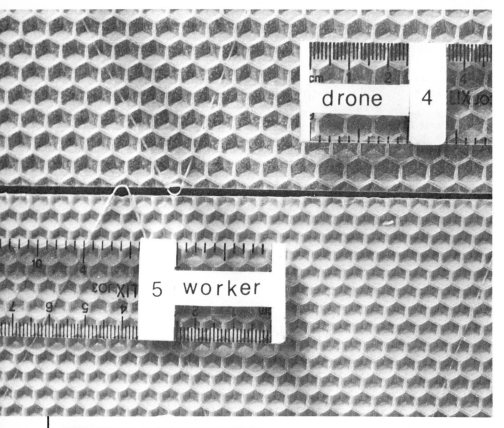

Above: *Worker and drone foundation*
Below: *A section through the finished cells that have been built from a sheet of foundation*

only in size. Worker cells have a diameter on all sides of 5.37 mm (⅕ in), and drone cells 6.91 mm (¼ in). The cells are all tilted upwards at an angle of 4-5 degrees to the vertical. The depth of the cells also differs. Worker and drone cells are 10-12 mm (⅝ in) deep, honey cells often up to 27-37 mm (1½ in) and queen cells 20-25 mm (1 in) deep. Queen cells can be used only once.

The cell form is an exact hexagonal

prism. They are closed at the bottom by three diamond-shaped planes, the obtuse angles of these surfaces forming the apex of a three-sided pyramid. Each wall and bottom is also a partition wall between two cells. The space is used to maximum advantage. Using the least possible amount of material, the bees achieve excellent structural stability and maximum heat conductivity.

Bee-houses, Hives in the Open, Bee Shelters

In Germany today, modern beekeeping techniques are gradually pushing the old-fashioned **bee-house** into decline. It has, however, many advantages for beekeepers who use rear-opening hives and those who are stationary rather than migratory.

The colonies can be kept in a small space: two colonies are arranged so that one is stacked above the other in two rows. The beekeeper sits down to deal with the bottom row and stands for the top row.

The hives are not so exposed to the weather and therefore have a longer life. The danger of robbing in late summer is also reduced. In winter the bee-house provides the bees with better protection from the cold.

The bee-house should be light and airy and should also double as a work-cum-storeroom. In Germany, taking over a bee-house from an older bee-keeper is obviously an advantage. In Britain a beginner would have to build his own since none are available commercially except at prohibitive custom-built prices. This would be in the order of £2,000 for 12 to 20 colonies.

The beekeeper also needs his own plot of land. In Germany, authorization for a bee-house is given only to experienced beekeepers. In other countries it would be wise to consult the local authority and take thought about possible nuisance to neighbours.

Top-opening hives are best kept **in the open**, preferably on the edge of a wood. The hives should be arranged with plenty of room between them: e.g. in groups of two with sufficient space between the groups to allow access to the colonies from three sides. The bottom stand, made from a framework of wooden beams, pallets or something similar, should be high enough to allow a three-storey hive to be comfortably manipulated from a standing position.

A bee-house at Merrist Wood; a traditional log hive is shown in the left foreground

It is a good idea for the alighting board for heavily laden returning bees to be an integral part of the base support, rather than being fixed to the bottom of the lower chamber. In summer, hives with sound roofs covered with roofing felt or sheet alloy should be sufficiently protected. In winter, an extra protective heat-insulating cover may be an advantage providing the air vents in the roof are not covered up.

The advantages of hives in the open are that they are simple and cheap, and outgoing and incoming bees do not create a disturbance. Set against this is the fact that the bees consume a great deal of food in winter and the hives have a life span of only about fifteen years even when they are weatherproofed. (Weatherproofing materials used must not be harmful to the bees.) In addition the beekeeper needs a small room to store equipment.

A **bee shelter** is a practical compromise between bee-houses and hives in the open and is designed specially for top-opening hives. The shelter is basically a shack with the roof covering extending to the ground. The front is made of boards with spaces in between and the shack is entered by a small door at the back.

The hives stand inside on wooden blocks. The shelter protects the hives from the weather and gives them a longer life. The roof is an advantage to the beekeeper because it provides shade in summer, and if it is raining he can continue working, although he should do this only in exceptional circumstances. Rain falling in can harm bees, brood and stores.

The Beekeeper's Tools

The basic piece of kit the beekeeper needs is a device to make smoke. In Germany, beekeepers sometimes use a pipe instead of a smoker with a bellows.

The Dathe pipe is held between the teeth. By blowing, the beekeeper can make as much smoke as he wants and direct it exactly where he wants it. The great advantage is that it leaves his hands free to work. However, the pipe has to be refilled after every colony and must be emptied and cleaned each time after use.

Many beekeepers find it difficult to keep the pipe in their mouth the whole time. Also, the bees find the shiny brass parts attractive and tend to fly towards them. However, the pipe is a useful tool in bee-houses and for rear-opening hives.

Beekeepers with top-opening hives normally use a smoker, a hand-operated device which has a pair of bellows and a metal cylinder to take the fuel. After the smoker is lit, the bellows are operated to supply the necessary air flow, but to use them the beekeeper has to stop what he is doing. The amount of smoke is difficult to control.

Suitable materials for burning might be cardboard egg cartons, dried pine cones, peat, dry rotten wood, chain sawdust, sackcloth or walnut tree leaves.

Even old hands often find it essential to have a **veil** to protect their heads and faces from stings. A black transparent material should be used

Above: *a range of smokers.*
Below: *Beekeepers well protected in hats and veils (left) and suit with hood (right)*

for the part in front of the face as black is easiest to see through. Most people prefer a veil fitting over a hat with a broad rim to keep the veil well away from the face.

The special beekeeping **boiler-suit** is made from a close-weave, white cotton material with a zip fastening at the front. The wrist and ankle bands are elasticated to keep bees out. If you do not want to buy a special suit, wear light-coloured clothes and stuff your trouser-legs inside thick socks.

You can also buy elasticated bee-keeping gloves made of soft leather with fabric gauntlets. Although a beginner may be inclined to wear gloves for fear of bee stings, it is not a good idea to feel obliged to do so because gloves can make you clumsy with the bees and actually incite them

Beekeepers' tools

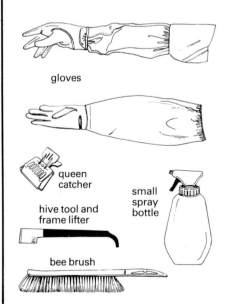

gloves

queen catcher

hive tool and frame lifter

small spray bottle

bee brush

to sting. You soon get used to the pain of stings. Some beekeepers rub oil of cloves into their hands; just a couple of drops send the bees away like magic.

Another vital piece of equipment is the **hive tool**, used to loosen the supers and combs which have become glued together and also to scrape propolis and wax off the frames. You can, if you need to, use a brush to push the bees off the comb, but on no account use one with stiff artificial bristles: a soft horsehair brush is ideal. A large goose feather, if you can get hold of one, is very useful for brushing the bees off harmlessly.

It is a good idea to make up a small tool box with spaces for all the utensils you need: pipe or smoker, fuel, matches or lighter, gloves, hive tool, brush or goose feather, hammer, small nails, secateurs and a screwdriver.

It is likewise useful to have a hand-held water syringe. You will also need a container of water with a cloth and towel to wipe up any spilt honey or sugared water and to wipe your own hands if they get sticky.

There are specialist beekeeping shops where you can buy all the equipment you need; they will be happy to answer any queries and give you tips on new beekeeping methods.

Finding Your Bees

One of your first questions will be 'How many colonies should I start with?' Obviously it is not advisable for a beginner to have too many colonies because the bees and the bee-

A light tray for carrying equipment

keeper will need to get to know each other slowly at first. It is best, therefore, to start with anything from two to five colonies.

One colony is not enough because, in your enthusiasm, you may be tempted to keep examining the bees. There is nothing worse for a bee colony than continually being torn apart. It disturbs their harmony and many bees, sometimes even the queen, may go missing.

If you start off with a few colonies you will be able to compare them and be sure of having at least one that is healthy and strong. Even if he or she inherits them, a beekeeper starting off with more than eight colonies will be severely overtaxed, and may quickly lose all interest in bee-keeping.

Your next question will probably be 'Where do I get my bees from?'

Not everyone has the good fortune to be able to take over bees and equipment from someone else. Normally you will have to buy your bees. A swarm is quite a good start: the bees build well, are peaceful and hardworking and need little looking after. They often raise a new queen without any problem, to become a first-class colony the next year.

Nowadays, however, it is not so easy to come upon swarms. Sometimes a friendly beekeeper may make a novice a present of one. In some places if you leave your address with the fire brigade or the local police, they will contact you when they discover a swarm. If you are lucky, you might even find one without an owner.

If you want to buy bees, look in the beekeeping journals for offers. Buy only healthy, disease-free colonies and preferably have them checked by an expert first. Do not buy a colony from an area in which there has been an outbreak of foulbrood in recent years.

In Germany, ownership of bees must be registered at the town hall. This is not necessary in Britain. If a British beekeeper joins a club affiliated to the BBKA (British Beekeepers' Association, see reference section) he will obtain insurance for a few hives, but if he increases his stocks he must consider his insurance position and if he makes profits he may be liable for tax.

WORKING THE COLONY

Inspecting Colonies

The golden rule is: inspect your bees only in sunny weather when there are no strong winds and when the temperature is above 15°C (59°F). It is also a good idea to wait until the field bees are out foraging. In rainy or stormy weather, or in the midday sun, leave your bees alone.

When preparing to inspect your bees, remember they do not like any strange smells. Sweat, perfume or strong-smelling aftershave may irritate them and make them sting! They are attracted to shiny objects and may fly directly at them. For this reason, take off your watch and any jewellery before going near them.

It is very important to work slowly and steadily without being tense or afraid. On no account should you panic, even if you are stung. Have all your equipment ready to hand: hive tool, tongs to lift the combs if you like to use them, brush, water bottle, and an empty hive box or stand for the frames.

Even if you are coolness personified, you should never open a hive, even for a quick look, without using the smoker. The bees recoil from the smoke and fill themselves full of honey. Their stomachs then swell up, making it difficult for them to use their stings.

Open the hive cautiously, puffing in a little smoke first. Take out the outer comb or drone frame and lean it on the front of the hive or in the empty box you have ready. After that you can use the hive tool to loosen the other combs and inspect them one after the other. It is important to replace the frames as they were, in order to disturb the bees as little as possible.

The following are the points to look for when you are inspecting the colony:

- Has the colony got a queen? If they are warbling they haven't. Can you see eggs, larvae and capped brood? If so, how many?
- How strong is the colony? How many combs are being covered? Does the colony need expanding or cutting down?
- Is the colony in swarming mood (important in May/July)?
- How is their pollen and honey supply?
- Should you give the bees additional food, put in a pollen or honey comb from another colony, or have they an adequate food supply from outside?
- Note all the details on a stock record card, keeping a separate sheet for each colony. You should also note down other points like the character of the colony, their readiness to sting, how they sit on the combs, whether they use a lot of propolis, how much honey they are yielding.

Inspecting for queen cells on the underside of frames in a double brood box

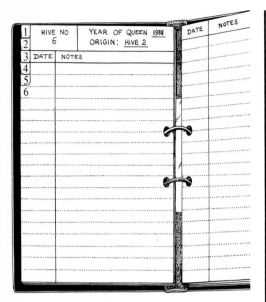

1	HIVE NO	YEAR OF QUEEN 1988	DATE	NOTES
2	6	ORIGIN: HIVE 2		
3	DATE	NOTES		
4				
5				
6				

A stock record card

Because it is very unpleasant for the bees to be inspected, the experienced beekeeper learns to look through the hive entrance to see how the colony is getting on. This saves work, too.

Colony Reproduction

Natural Reproduction (Swarming)

A bee colony, as we have seen, is only viable as a complete entity. Colonies are constantly being killed off by disease or lack of food as a result of severe weather, so it is interesting to find out how a new colony is created.

A new colony is formed quite simply by division – the process of swarming in which the old queen leaves the colony with a proportion of the bees.

The Swarming Process

Early summer, the climax of blossom time in nature, also marks the climax of the life of the bee colony. The rich honey flow from fruit blossoms, dandelions, oil-seed rape and other plants provides the bees with all they need to rear brood and do their building work. There are many factors, however, which may make them swarm: bad weather which stops them foraging; a superabundance of brood food from the many young bees; lack of space; an ageing queen; the intense heat of the midday sun; or lack of food and water.

The worker bees make queen cup cells on the edges or underside of the frame and the queen lays eggs in them. The queen's egg-laying activity then begins to decline. The worker bees supply her with less food and start chasing her round the combs.

As her ovaries shrink and she loses weight, she regains the capacity to fly. In the colony building work and foraging is cut back and the bees tend to hang around the hive entrance in a big cluster. The queen cells, sometimes up to twenty in number, are still cared for.

Nine days after the eggs have been laid in the first queen cells comes the moment when, weather permitting, the colony divides. Well over a third of the bees fill their stomachs with honey for the journey. At around midday there is a loud buzzing as the prime swarm emerges and makes zigzag movements across the sky.

Gradually they begin to hover in a huge cloud which forms into a thick cluster and comes to rest on a tree, roof beam or some other similar place. Up to 20,000 bees can collect like this in the vicinity of their old home before moving to their new one – which may be miles away.

Scout bees are sent out to look for suitable accommodation; on their return they perform a dance for the other bees, pitching their level of excitement in proportion to the desirability of the home they have found.

Often the swarm will move into a hollow tree, a hole in a wall or somewhere similar. Only if the queen goes missing will the bees return to their old colony.

Catching a Swarm

Swarming can be as tense a time for the beekeeper as it is for the bees! A new beekeeper will probably be filled with a mixture of apprehension and excitement when one of his colonies swarms.

The first thing to do when this happens is to wait patiently until all the bees have settled in one spot in a cluster. Collect together a syringe with water, a veil, possibly a ladder and something to catch the swarm in (a bag, straw skep or empty hive box with a floor and lid).

If only a few bees are flying around the cluster, you can spray it with a fine mist of water. This discourages the bees on the outside from flying up when the swarm is caught. Now comes the big moment when you actually catch the swarm. Hold the container in one hand under the cluster, as near to it as possible, and bring the other hand down sharply on the branch, or whatever, to knock as many as possible into the container.

Swarming can reduce the parent colony to less than a third of its size. In the prime swarm (1) the old queen mother leaves, taking more than a third of the bees. In two afterswarms (2,3), virgin queens depart, each with about a sixth of the original population. The number of afterswarms can vary from none to four or five.

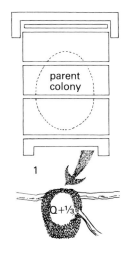

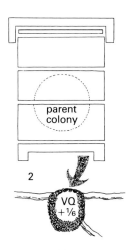

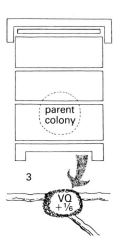

If you have managed to catch most of the bees, put the container down on the ground in the shade beneath the place where the cluster has been hanging and open the flight hole, or prop the container up to allow the rest of the bees in.

You will soon see if you have been successful. If you have caught the queen, all the other loose bees will fly into your container. If, however, she is not there, the bees will all fly out into a cluster again and you will have to repeat the whole operation.

If the swarm alights on a thin branch, simply cut off the branch together with the cluster and shake it firmly into the container. Sometimes swarms fly onto high branches, and

A swarm catching bag. The traditional method of catching a high swarm, using a small straw skep supported by a pitchfork, is shown on the left

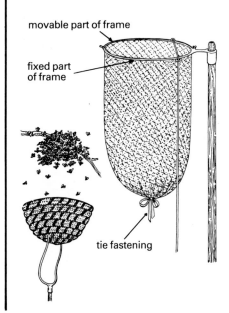

movable part of frame

fixed part of frame

tie fastening

this is where the special catching sack comes in handy. It can be fixed to the end of a long pole, allowing the bee-keeper to reach up to the cluster. The bag has a metal rim round the top, half of which is movable. By pulling a string, the movable half is folded over onto the other half.

A swarm will not always form a tight cluster. Sometimes the bees spread themselves out, along the trunk of a tree for instance. In this case, spray them gently with a fine water mist and use a brush to sweep them into your box.

In doing this, you will be able to see how remarkable a swarm is. The bees are linked lightly together and are perfectly easy to sweep up. Providing you work steadily, they won't be the least bit interested in you, being completely taken up with themselves.

Swarms which have clustered in very inaccessible places – in large towns, for instance – are often treated as an emergency, and fire services or local authorities will be called on to retrieve or destroy them.

What to do with the Swarm after you have caught it

Once you have caught your swarm, put it in a dark place for one or two days. If it is a swarm from another area, you would be advised to keep it locked up for three or four days just in case the bees have brought any diseases with them. After this period of darkness, transfer the swarm to a clean hive.

The rule of thumb is to allow the swarm one frame per 227 g (½ lb) of swarm weight. You should also insert a comb which has already been drawn out so that the queen can start laying eggs immediately.

Sometimes the bees will swarm again. To prevent this, give them some uncapped brood comb – they will not desert the brood. There is no problem with placing the swarm near its original colony, because after swarming the bees will have forgotten about it.

When there is no honey flow, or in periods of bad weather, you will need to feed the new colony about four days after swarming. You should do this only in the evening, to prevent robber bees from getting in. The bees will need about ½ litre (1 pt) of a honey/sugar solution for three days.

What to do with Weak Swarms

Small swarms which occupy only a few combs can be built up gradually by the introduction of surplus capped brood combs from other colonies. In this way they can be made sufficiently strong to survive the winter. Alternatively, you can unite swarms. The rule in this case is to make sure that a prime swarm is united only with another prime swarm and an after-swarm only with another after-swarm. Otherwise you may lose a lot of bees.

A prime swarm is the first swarm with the old queen. Since the queen has already mated, she will soon be able to start laying eggs. The after-swarm occurs when a virgin queen in the parent stock moves out with another section of bees about seven to eight days after the prime swarm.

What happens to the Parent Stock?

To find out which colony has swarmed, you can do a 'flour test'. In the evening, take a couple of bees from the swarm you have caught, put them in a glass with a little flour and release them in front of the hives. The hive into which the floury bees fly will be the one that has swarmed.

Generally a colony left behind after a swarm will need to be cut down. All but one or two of the queen cells should be taken out. After about a fortnight, check to see whether a queen has emerged and started to lay.

Once you know for sure which colony the swarm has come from, you can put the swarm in a new hive in its old position and the parent colony nearby. The field bees from the parent colony then fly to the new hive, so that the swarm starts operating to the full and producing honey. The parent stock can be used to form a young colony or be united with a weak colony.

Swarms used to be highly sought after, particularly those found early in the year. Heathland beekeepers used actively to encourage swarming. Because heather blooms late, the parent stocks had time to recuperate and the swarms were also able to produce a crop of honey. After-swarms were also highly prized.

Nowadays beekeepers tend to go in for intensive swarm prevention measures. If a beekeeper has a job, the chances are that his place of work, home and apiary will all be far apart, leaving him little time in the summer to go off catching swarms.

Another disadvantage of swarming is that the bees do not work as hard beforehand and they take honey with them, leaving less honey for the beekeeper.

Possible Ways of Counteracting Swarming

These are:
- giving the bees enough space at the appropriate time;
- giving them enough frames to build on;
- making sure the queens are young;
- keeping them out of the midday sun.

The important thing is to prevent them from getting into the swarming mood. Once a colony has decided to swarm, any preventive measures are likely to be too late. Generally speaking, about 10 per cent of colonies will swarm each year. There are some years when the bees give no thought to swarming and other years when nothing will stop them.

In hives with supers it is often possible to prevent swarming by the use of a dividing board which is described below.

In the swarming period of May/June, colonies need to be checked for swarming at nine-day intervals. The points to look for are whether the colony is in swarming mood (have they stopped building and foraging? Has the queen been laying fewer eggs?) and whether there are any queen cells.

Lift off the supers and look at the

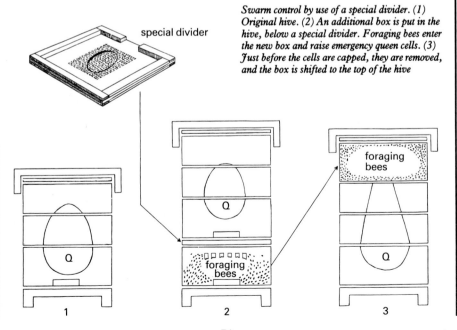

special divider

Swarm control by use of a special divider. (1) Original hive. (2) An additional box is put in the hive, below a special divider. Foraging bees enter the new box and raise emergency queen cells. (3) Just before the cells are capped, they are removed, and the box is shifted to the top of the hive

foraging bees

foraging bees

1   2   3

undersides of the combs. Since the queen cups are usually constructed on the undersides of the frames on the top brood chamber, they are easy to see. But use the smoker first to calm the bees. As soon as you find a queen cell with an egg in it, you must do something to prevent swarming.

Preventing Swarms without Creating a Young Colony

Perforated dividing board The forager bees are first separated from the parent stock in swarming mood by using a dividing board, and then reunited. There are two steps:

1 Take off all the boxes and, on the floor board, place a new chamber with two brood combs in the middle, frames of foundation next to it, and to the left and right a honey comb for food.

Place a special perforated dividing board through which the bees cannot pass with a flight hole above it on top of this chamber. After taking out all the queen cells, replace the colony in the original order. The queen is now in the original hive above the special dividing board.

The foragers return to the hive through the old entrance at the bottom, leaving plenty of room on top of the special dividing board and a dearth of field bees. This causes a lot of the young bees to start flying and the queen to start laying eggs again with renewed vigour.

It is important when doing this to ensure the special dividing board is inserted before 3 p.m. while the bees are out foraging. In the bottom chamber, the field bees will build emergency queen cells on the two brood combs.

2 This part of the operation has to be carried out nine days afterwards, before the emergency queen cells are capped. Remove the special dividing board and replace the hive in its original order on the floor board. Then remove all the queen cells from the brood chamber, with the field bees, and put it on top.

Usually the frames of foundation will have been drawn out and filled with honey. The best idea then is to hang the two brood combs in the bottom chamber below the field bees and the honey-filled side combs above. In many cases this will be enough to overcome the swarming instinct and keep the whole colony foraging.

Artificial Reproduction (How to Create a Young Colony or Large Nucleus)

There are two main ways of preventing swarming and at the same time creating a young colony: (1) artificial reproduction by making a nucleus; and (2) creating a young colony of bees only and adding an impregnated queen; or by creating a young colony of bees and brood comb and adding a capped queen cell or a queen.

First method Before a strong colony gets into a swarming mood, it is possible to create an 'artificial swarm' or large nucleus. To do this, sweep about eight to ten thousand (about 2 kg (4½ lb)) in weight) into a suitable box. The bees should be made to pass through a queen excluder, because then if the queen is amongst them, she can be found on the mesh and returned to her colony.

The bees should then be transferred to a hive with two foundation frames on the outside and three frames in the middle which have had all eggs and brood cells cut out. The remaining space should be filled with foundation frames. Introduce an impregnated queen into the hive in a cage sealed with candy. Keep the new colony in a cool, dark place for one or two days, giving it some additional food. Because inserting a new queen is risky, it is more likely to be successful if there are no open brood cells which would give the bees the opportunity to create further queen cells of their own.

It is important to locate the nucleus swarm at least 3 km (2 miles) away from the parent colony to prevent the young bees from returning.

Second method Place three combs, with as much capped brood as possible, with bees from five or six brood combs, in a new hive without a queen, together with a comb of food and some water.

The new colony should then be moved to the nursing apiary. After eight to ten days, remove the new queen cells and replace them with a queen cell from a good colony on the point of emerging, or a queen.

About fourteen days later the new queen will have hatched and be laying eggs. At this stage a Varroa check can be carried out (see pages 126-7) as all the old brood will have hatched and only eggs and open brood will be present.

There are two variations to this method.

a) A colony on the point of swarming can be moved during foraging weather to a new position within the old foraging area. This should be done before midday. Place a new hive, painted the same colour as the original hive, in the old position.

The new hive should have two open brood combs in the centre, two combs with food on the outside and foundation frames in between. All the field bees from the colony will fly into this new hive. In this way the colony is 'bled' of foragers and will usually lose its swarming impulse.

The new colony will build emergency cells for a new queen on the brood comb. These must be removed after nine days. If the brood comb has come from good stock, one or two cells can be left. Otherwise, another queen cell, ready to hatch, or a young queen should be inserted. However, it is important to note that when adding a queen cell, the new colony must be removed to the permanent site after fourteen days at the latest. By this time all the old brood will have hatched, allowing successful Varroa treatment.

If an already impregnated queen is added, the new colony must be removed to another apiary after nine days at the latest. The Varroa treatment should be carried out before the queen is introduced.

b) The other variation involves 'drawing off' bees from strong colo-

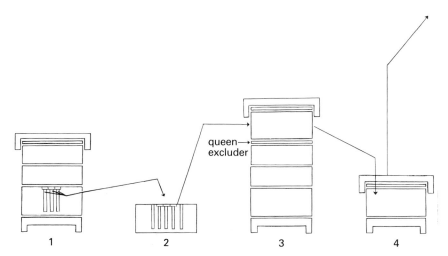

Swarm control by drawing off younger bees. (1) Brood combs are removed from a hive and (2) placed in a separate brood box; (3) new box is placed on top of a colony that is in danger of swarming. (4) One day later, it is removed to another apiary, taking with it its population of younger bees

queen→
excluder

1   2   3   4

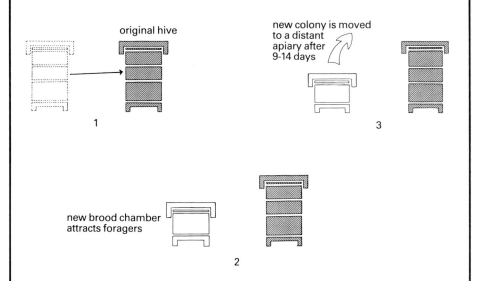

original hive

new colony is moved
to a distant
apiary after
9-14 days

1

new brood chamber
attracts foragers

2

3

Swarm control by drawing off older foraging bees. (1) When the colony is on the point of swarming it is moved a short distance. (2) A new brood chamber is placed on its original site. Foraging bees return to this chamber, which can then be removed to a permanent site as far away as possible (3)

nies to prevent swarming. Around about May, before any queen cells indicating swarming have been built, place two or three brood combs, light brown in colour and sprayed with sugar water, in the brood chamber. These should be marked with drawing pins for identification purposes.

Six or seven days later remove the combs with their young brood, regardless of the weather. Brush the bees off gently to ensure no larvae or eggs fall out, and place the combs, together with two combs of food and foundation frames, in an empty box.

Put the brood box on top of a colony in danger of swarming, with an excluder in between. The young brood will draw nurse bees in to look after it. One day later the box can be removed, a floor and crown board added, and placed in another apiary. After eight days the queen cells can be cut out and an impregnated queen introduced.

How to Introduce a Queen Cell or Queen

An approximately fourteen-day-old queen cell, from a good colony which has swarm cells, can be introduced into a young queenless colony. You must remember, however, that the day before she hatches, the queen will be extremely sensitive. The queen cell should be fixed with wax to the middle of the centre comb in the new colony.

Introducing a new queen which you have purchased must also be carried out very painstakingly. The queen should be placed in a cage sealed with candy and hung between

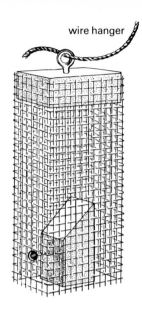

A queen cage

the brood combs. The bees eat through the candy and are then generally ready to accept the queen. However, there should be no other queen or queen cells in the colony.

Looking after a Young Colony or Nucleus

If a ripe queen cell is introduced, it will take about three days for the queen to hatch and about fourteen days for her to start laying eggs. Afterwards you must ensure that the colony always has enough space and enough to eat. Generally speaking, you will need to give them additional food – about 2 kg (4½ lb) of honey or sugar.

If the queen goes missing on her nuptial flight, new queen cells can be

A virgin queen emerging from her cell

added again or, if it is late in the year, the colony can be united with another young colony. The average loss rate for new colonies is at least 10 per cent.

After a further three weeks, the colony should be inspected again. You should already have brood and young bees from the new queen – if you find a closed brood nest, you will know you have a good queen. If everything is in order, you can expand the colony and give them additional liquid nourishment. To ensure they are strong enough by late autumn to survive the winter, it is best to start feeding the colony early.

Feeding

Bees absorb sugary solutions from nectar which they process by adding enzymes and drawing off water. The same happens to the sugar the bee-keeper gives to them and this, too, should be in solution.

Since bees can easily drown in liquids, you should make sure the feeder you use has some way of allowing the bees to get a firm hold, perhaps by putting moss or woodwool in the base. There are various feeding systems on the market now: contact feeders which are placed over the hive, frame feeders, indirect plastic or tinplate feeders, Miller feeders.

Feeding in Spring and Summer

If food supplies are low in the spring, the beekeeper can take a comb full of honey from a hive with an abundant food store and hang it directly next to the brood nest in the hungry colony. Such bees can also be given leftover honey skimmings mixed with wax cappings. The important thing is to try to give them only invert sugar in the spring.

Candy can be given to the bees later in the year when the weather is warm and the bees can take in water without harm. Candy is useful for feeding young colonies and stimulating commercial colonies for brood rearing.

The recipe for candy is: four parts icing sugar to one part warmed honey, mixed well together. The firm paste can be placed on the crown board or on top of the frames, where the bees quickly get at it. About ½ kg (1 lb) per colony is enough for around ten days. The candy can be kept for a long time in sealed containers. It is also made commercially.

Forty days prior to what looks like being a good honey flow, bees may be fed to incite them to brood activity. (It takes forty days for an egg to develop into a forager bee.) In this case the bees are best fed with sugared water in a ratio of 1 kg to 1 litre (2 lb to 2 pt), or sugar paste if the weather is good.

Winter Feeding

In winter the bees should be given sugary syrup. Dissolve three parts sugar (ordinary household sugar) in two parts hot water, stirring the mixture until it is clear. A simpler formula is to heat up 1 litre (2 pt) of water with 2 kg (4 lb) of sugar. It is a good idea to give the solution to the bees while warm.

About 750 g (26 oz) sugar and 500 g

Candy laid on the crown board for emergency feeding in winter

(17 fl oz) water give 1 litre (2 pt) of syrup. If you want to pamper your bees, add a pinch of cooking salt (but not after the bees have been foraging in forests) and an infusion of tea made from a mixture of camomile, thyme, yarrow, nettles, dandelion, valerian and oak bark (10 g [⅓ oz] of leaves to 100 litres [200 pt] of liquid).

The bees need enough food to take them through to March/April. The amount they need depends on how you keep them. Bees kept in bee-houses are usually in a more restricted space during the winter and tend to be better protected from the cold because the hives are closer together and are insulated with felt inside. Bees in these hives require about 10 kg (22 lb) of food per colony.

In top-opening hives the bees generally overwinter in either one or two hive-boxes, depending on the strength of the colony. A colony in a single box will need at least 10 kg (22 lb) and one in two boxes at least 12-15 kg (26-33 lb).

Feeding is more time-consuming with rear-opening hives, because the feeders hold only about ¾ litre (1½ pt) of liquid. The feeders need to be filled each evening for about two weeks. In hives with supers larger amounts of food can be given at one time. The rate the bees consume the food depends on the temperature. Generally speaking, they will be ready for a second helping within the week.

Before giving the bees supplementary food, remaining stores of pure honeydew honey should be removed. Honeydew honey puts a severe strain on the digestive organs in winter. If the bee does not get an opportunity to leave the hive to defecate, it runs the

81

Bees foraging for nectar and pollen on heather, willow and clematis

risk of developing dysentery. This is not a common problem in the UK or in districts anywhere where little honeydew is gathered in normal seasons.

If you can begin winter feeding in plenty of time while the weather is still quite warm, you can use candy, but it is more expensive than syrup.

Migratory Beekeeping

Migratory beekeeping has been practised throughout the ages. The ancient Egyptians transported their bees on ships along the Nile. Nowadays commercial beekeeping without moving bees around would be completely unthinkable.

Migratory beekeeping is the practice of moving colonies to a place which provides the bees with a better source of forage. Even an amateur beekeeper should familiarize himself with transporting techniques because they involve very basic principles of apiculture.

In many places it is difficult to ensure that the bees will have access to the pollen and nectar, or honeydew they need throughout the growing period. Modern agricultural methods have made it very difficult for bees to find sustenance in intensively farmed areas.

Meadows are often mown early and all the weeds killed off with herbicides. There will often be periods when there are gaps in the honey flow. If this happens, a beekeeper will often have to move his bees a great distance to find another source of forage.

Woods (particularly pine and fir forests) are favourite places for migratory beekeeping. Since they don't have a honey flow every year, however, beekeepers have to move their hives if they want a crop of much sought-after wild honey from woodland areas.

Some Rules about Migratory Beekeeping

You cannot always put your bees just wherever you like. In Germany, for example, before moving bees outside the area of your local association, you have to have them checked by a bee expert. German beekeepers are allowed to move healthy colonies only – a precaution against the spread of diseases, as the danger of an infection being transmitted is particularly great

Hive stands

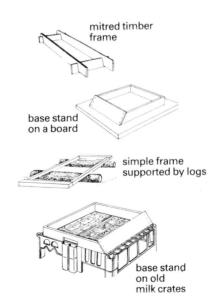

mitred timber frame

base stand on a board

simple frame supported by logs

base stand on old milk crates

in areas with a good honey flow.

As a migratory beekeeper you are responsible for finding your own site, but do spare a thought for the local beekeepers. There may be regulations that forbid you to set up your hives in the vicinity of their hives, or in the foraging radius of crops acknowledged as foraging ground for local bees.

Decide on your site in collaboration with the local person in charge and agree on a rent with the owner of the land. You can make this proportional to your honey crop.

The site should be easily accessible by car and near to the honey crop. It is best to choose a site on the edge of woodland, even when using a flower crop. Often you will need to get the place ready first: clearing undergrowth and preparing a base for the hives.

You may wish to construct a simple wooden base, or simply use a piece of Chipboard, or an old car tyre. Alternatively you can even use a piece of carpeting or roof felt as flooring, protruding about 30 cm (1 ft) in front of the hive. This will keep grass and water from getting in the hive entrance.

A bituminous corrugated roof held down with stones can be used to cover hives that have been removed from bee-houses. In an emergency, black plastic bags are a good standby. You may also need to set up a source of water.

How you get the bees ready for transport will depend on the type of hive you use. The important thing is to make sure the front of the hive is properly sealed so that the bees can-not get out, but still have plenty of air. If you have rear-opening hives, you can simply remove part of the door at the back and replace it with perforated netting to allow air in. Hives with supers can have their entrances blocked with furnishing foam and be fitted with locking devices. Otherwise they should be secured with a belt or coppered staples, hammered in obliquely.

Do not put new frames into a hive directly before moving the colony. They will not yet be covered in bee glue and will tend to slide around, squashing the bees. Only move strong colonies with a food supply sufficient to last for two weeks. If their food supply is not adequate, feed them three days before setting off.

When you are ready to move the bees, the entrance to the hive should be shut off at night once all the bees have returned. You can start transporting the bees either immediately or early next morning. If possible, arrange to be at your new location before the bees want to go out foraging.

It is vital to ensure that they have sufficient ventilation on the journey or the biggest, and therefore best, colonies might suffocate. If the bees are being moved a long distance, place some moss soaked in water near them to keep them cool.

Use a suitable vehicle for transporting them: a car with a trailer, or a delivery van or lorry. Load and unload the bees as gently and steadily as possible: a barrow is useful. Because of the danger of accidents, you should on no account travel alone with bees.

A bucket of caulking material or

Above: *Taking a swarm in a straw skep*

Left: *Swarm emerging from a hive*

Right: *Swarm pitched on a tree*

Securing a hive with a belt for transit

strong, wide sticky tape comes in handy to seal any cracks in the hives. You may also need a hammer, pliers and nails.

The journey should be as smooth as possible – avoid making any stops if you can. If you have to stop, keep the engine running in order to keep the bees quiet. If you have to take a long break, particularly in hot weather, the bees must be unloaded and allowed to fly. The journey should then be continued in the evening.

When you reach your destination, do not open the hive entrances until all the colonies are on their stands. In fact it is a good idea to wait even longer. Put on your veil and open the hive entrances, puffing in a little smoke first.

Here is a checklist for all you require before transportation:

- vehicle: with sufficient fuel and the vehicle's documents;
- belts to secure the hives;
- beekeeping tools: smoker, fuel, matches, hive tool;
- other tools: hammer, pliers, nails, broad adhesive tape;
- protective clothing, veil, gloves, strong shoes;
- water container and spray;
- torch;
- first-aid kit;
- health certificates (where required).

The Honey Crop

When the beekeeper finally sees capped honeycombs hanging in the hives, his heart will probably beat faster, because this is where the 'sweetest' part of his work begins. The honeycombs must not be removed too soon, however, because otherwise the honey will have too high a water content and may ferment.

The water content should be below 20 per cent. The right time to remove the comb is when the honey is about two-thirds capped. But there may be times when the honey will be ripe even before it is capped.

Honey from oil-seed rape, wild mustard and larch hardens quickly in the comb. To test if the honey is

Removing a frame

ready, remove any bees from the comb, hang it vertically and give it a firm shake. If no honey spurts out of the comb, it is ready to be extracted.

If you wait too long to remove it, the bees will be satiated and will not feel inclined to work so diligently. Also, some types of honey, such as larch and oil-seed rape, can become so cement-like that it is impossible to get them out of the comb.

It is advisable to have a helper when it comes to removing the honey. The quicker the job is done, the less likelihood there is of the bees stinging or of robber bees getting into the hive. Everything should be well prepared beforehand: tools, water spray, transporting boxes or empty hive boxes with cover boards.

Remove the top of the hive, using the smoker, and pull the cover back halfway. Remove the ripe honeycombs one after the other, knocking and brushing off any bees, and hang them in an empty super.

Cover the hive again immediately to avoid disturbing the bees too much. Generally the honeycombs will be on the top and at the sides. If brood combs appear in the middle, put them on one side with the brood-box.

Repeat the operation with the rest of the supers. Once you have finished, put any brood frames in a super over the brood box. For the time being, the top super can be replaced empty if you have no other empty combs available.

As soon as the honey has been extracted from the combs, the frames should be sprayed with water and, either the same evening or next morning, hung back in the hive. If the

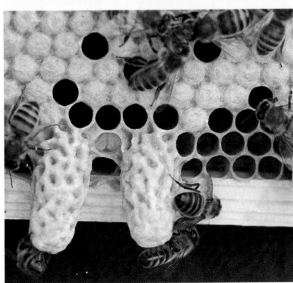

Above: *A superb frame from the centre of a brood nest showing sealed brood and an arch of capped honey*

Left: *Swarm cells on the bottom of a comb*

Right: *A queen cell cut open to display the pupa and royal jelly it contains*

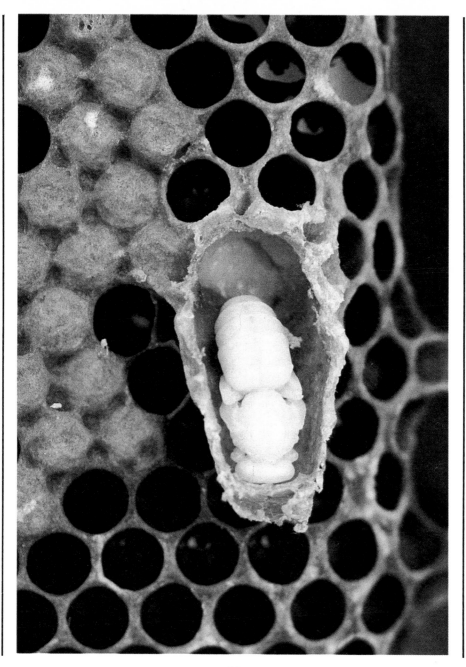

Well-filled shallow frames of honey

colony no longer needs so much room, the top super can simply be left off.

Extracting the honey should be carried out in a bee-proof room. Absolute cleanliness is essential. Since the combs need to be spun while they are still warm from the hive, it is a good idea to put them in a heated room or to cover them with film and put them in the sun whilst awaiting spinning.

Uncapping

The first task is to uncap the combs. A double-bottomed metal tray with a wire insert is useful both as an uncapping table and as a container for the wax cappings. In some countries uncapping is done using a broad, heavy-duty fork with its tines bent slightly upwards; in others, including Britain, a knife is more usual.

The fork (if used) is inserted just below the cappings at the bottom of the comb and moved upwards to remove the fine layer of wax.

A similar upwards sawing motion is employed when using a hot carving knife blade or a special electrically heated uncapping knife. Allow the cappings to drop into the tray and try to leave a straight face on the comb.

Uncapping with knife and tray

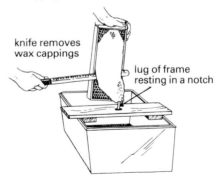

knife removes
wax cappings

lug of frame
resting in a notch

It is important to ensure that when cleaning sticky equipment, no water gets into the honey as this would spoil its quality.

The extraction process itself depends on the type of extractor used. In principle, the uncapped combs are placed in a rotating steel cage with the weight distributed roughly evenly.

The extractor can either be hand-operated or electrically powered. Do not use top speed immediately. If you are using a tangential extractor, spin one side of the comb gently first, then stop the extractor, turn each comb and continue to spin until the second side is empty. Turn the combs again and the side of the comb spun first can then be emptied completely.

New combs in which no brood has been raised need to be handled very carefully because they break easily. These may need to be turned again in the extracting process. Commercial beekeepers often use self-turning extractors in which the combs are turned automatically in the opposite direction.

Radial extractors allow frames to be inserted like the spokes of a wheel, and they do not need to be turned at all.

The honey runs down the inside of the extractor and collects at the bottom. The 'liquid gold' can then be drawn off through a tap and passed through large- and small-mesh filters to strain off any wax debris, into a storage tank with an extractor tap.

The way the honey runs at room temperature tells the beekeeper something about its water content. If it forms peaks as it runs out, it is ripe and ready for storage. If it forms a

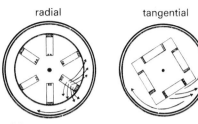

radial   tangential

A honey extractor works by whirling the comb in a centrifuge, which can be based on either of the principles shown here

A small honey extractor

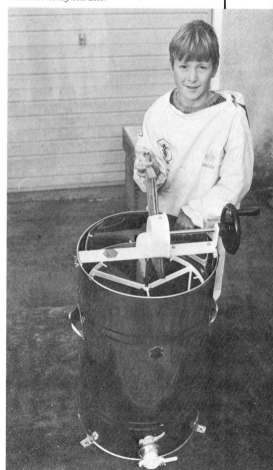

Above: *A mouse caught in a hive which it has destroyed*

Far right: *A wild colony of bees exposed in the cavity behind the clapboards of an old barn*

Right: *Honeycombs ruined by wax moth larvae*

hollow, however, there is still too much water in the honey and it could ferment.

Processing

Honey starts to set at different rates depending on the proportions of fructose and glucose. The higher the glucose content, the faster it sets. Floral honeys tend to set particularly quickly – not that anything is wrong with crystallized honey. Wild honey, acacia honey and honey from sweet chestnuts remains clear longer.

The beekeeper must process the honey to ensure it sets evenly. After extraction it should be left to stand in a warm place in the storage tank for a few days, and stirred well twice a day for ten minutes, using a three-edged wooden stick.

The container should on no account be made of untreated iron or zinc and should be covered to stop the honey absorbing foreign smells. All the wax particles and scum which collect at the top should be skimmed off to leave the honey clear.

Storage

Honey should be stored only in large containers made either of high-grade steel or plastic suitable for holding foodstuffs. It should be kept in a dry place out of direct sunlight, at a temperature below 18°C (64°F). Containers must be airtight to keep out water and foreign smells.

When liquifying solid honey, the temperature should not be allowed to go above 40°C (104°F), or the honey may lose its flavour. It can be heated carefully in a water bath or a warm cabinet (converted refrigerator). There are also some commercially made devices for the purpose.

Sales

Ideally the honey should be transferred to jars or containers shortly before being sold. Use high-quality

Honey warming box. Sizes vary but for convenience the inner box should be large enough to take two 12.7 kg (28 lb) honey tins

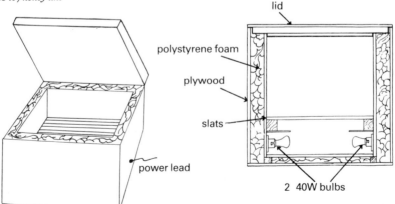

polystyrene foam

plywood

slats

lid

power lead

2 40W bulbs

Honey storage containers with strainers

containers and ensure that the weight is correct.

There is no reason why you should not ask a reasonable price for a high-quality product. Just because you might have had a good honey crop, do not be tempted to mark the price down – there are sure to be bad years to follow.

You can also sell the splendid pure, pale-coloured honeycomb cappings bottled in a screw-topped jar. Eaten daily, wax is a good remedy for colds, flu and sore throats. First remove the top layer of dry cappings from a full uncapping tray. Leave the customer the bottom layer of cappings soaked in honey which has dripped down from those above. If you cannot sell it, you can put it with the honey scrapings and use it to feed the bees next spring.

Honey is named according to what

Squat honey jars and other containers in regular use

it is made from. Honey from nectar is floral honey, wild honey is honey made by the bees from honeydew from conifers, and leaf honey is made from honeydew from deciduous trees.

Floral honey comes from the nectar of a variety of plant types and contains a little pollen; it is named according to the dominant nectar. There is, for instance, the thin yellow fruit blossom honey, the slightly bitter-tasting dandelion honey, the sweet whitish crystallized honey from oil-seed rape and the dark yellow, rather tart-smelling heather honey.

Honeydew has a lower pollen and acid content, but a higher mineral content. Honey from pine trees and spruces is much sought after and very expensive.

Rendering Wax

The beekeeper can collect wax from burr and brace comb removed from the hives, from cut-out drone cells and from old combs. As a basic rule, no hive should contain brood combs more than three years old. Freshly produced virgin wax is pure white, but darkens in the course of time. The pupal skins which remain in the cells every time a bee hatches gradually restrict the room in the cells. If the combs are not renewed every three years, the bees produced in them will be noticeably smaller.

Trials with artificial wax combs have proved that there is no substitute for the real thing. The beekeeper should therefore keep a bucket, with a lid, for wax debris to be returned to the bees for processing. After all, the bees have to secrete 1,250,000 wax scales to produce just 1 kg (2 lb) of pure wax!

There are various ways of extracting pure wax, but unfortunately the simple ways are extremely time-consuming and inefficient.

Because the necessary equipment for doing the job efficiently is very expensive, a small-scale beekeeper might be advised to sell his old combs or exchange them for foundation. In Germany, about 3 kg (6½ lb) of old comb can be exchanged for 1 kg (2 lb) of foundation, with a handling charge on top.

In most countries beekeeping

Wax scales that have dropped to a hive floor – 1,250,000 per kilo

98

A solar wax extractor

equipment dealers will buy wax rendered into blocks or make an appropriate reduction in price for new foundation in exchange.

Old combs should be packed for selling in a moth-tight container.

Some beekeeping associations keep a steam wax extractor which members can use. The old wax is boiled for about two hours, melts and is forced out with the condensed water through a tap into a bucket. It then congeals into a solid mass on the surface of the water. This process produces a decent yield and saves the trouble of cutting

the combs out of the frames and later rewiring them.

The best method, however, is to employ a solar wax extractor which uses energy from the sun. This is cheap, can be left to work unattended and is also suitable for processing small amounts of wax. It should have room to hold at least one frame, but the bigger it is the more convenient it is to use.

The solar extractor, which can easily be made at home, consists of a box with a dark tray tilted at an oblique angle (made from slate, enamel or tin plate) and a container to catch the wax at the front. The whole thing is sealed with a sheet of glass to make it airtight. If this can be double-glazed, so much the better.

The melting surface should be at right-angles to the rays of the sun. The heat accumulates under the pane of glass and once the temperature has reached 63°C (145°F), the wax melts. It then flows down through a spout into the collecting dish, where it sets at a temperature below 60°C (140°F).

Young wax (from drone cells and burr and brace combs) melts without residue, but the yield from older combs is relatively low.

Preparing the Frames

Used frames, with the wax removed, should be disinfected either by dipping in a 5 per cent soda solution (wear protective clothing, goggles and gloves) or putting through the flame of a camping gas burner or blowlamp. If you use soda for disinfecting, it will also remove all the remaining wax, whereas using a flame leaves a thin layer of wax and propolis behind. Frames dipped in soda need to be wired with high-grade steel.

Wiring is essential for the large frames which are now common, because it holds the foundation securely and makes the combs more rigid for the extraction process.

Use a punch to stamp holes in the sides of the frame through which the wires pass. The punching operation rims the hole with a brass eyelet which stops the wire cutting into the wood and makes it easier to thread. Large-scale beekeepers drill the holes with a boring machine; you can drill and press the eyelets in by hand. The position of the holes can be marked with a template.

A wired frame (horizontal wiring)

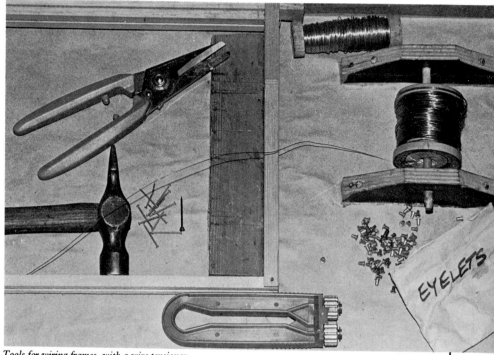

Tools for wiring frames, with a wire tensioner

The wiring can be done in one of two ways: vertically (with the wires running parallel to the upright sides of the frame) or horizontally (with the wires parallel to the top and bottom bars).

For horizontal wiring it is best to use bright galvanized or 26 swg monel wire. The outer wires should not be more than 12 mm (½ in) from the edge of the frame to prevent the foundation from curling up.

Advantages:
- The wire can be more easily stretched between the shorter sides than the more flexible, longer top and bottom bars.

- There are no wires in the way when removing wax remains from the frames.

Disadvantages:
- If hardwood is used for the sides of the frame a drill must be used, not a punch.

For vertical wiring (more commonly used on certain German frames) a thicker wire (0.5 mm) is preferable.

Advantages:
- The holes can be made by hand.

Disadvantages:
- The top and bottom bars might bend
- The wires get in the way when scraping off the wax.

Thread the wire through the eyelets (for a British Standard frame: four times for horizontal wiring). The wire must be tensioned correctly to avoid the foundation crumpling. It must not, however, be so taut as to cause the wooden bars to bend. Fasten the ends with two small flat-headed nails hammered into the wood with a small hammer near the holes on the outside of the frame.

As far as foundation is concerned, you can either make it yourself if you have the right equipment, or buy it from a specialist shop. You will need to know the exact measurements of your frames. The foundation sheet should be slightly smaller than the aperture of the frame because it expands slightly in the warm brood nest and would tend to bulge if it were exactly the right size.

The embedding of the wires into the wax is generally carried out in spring, using an electrical current reduced by a transformer, or a battery charger to 8-12 volts. The transformer contacts touch the ends of the wire and the nail heads; as they heat up, the wax begins to melt and the wires become embedded in the foundation. The electric current must be switched off in good time or the foundation will melt right through. As the wax cools, it becomes firmly fixed to the wires.

Points to note are:
- The sheets of foundation should not be embedded cold. To ensure they are warm, either lay them separately in the sun for a while, store them at room temperature, or put them near an oven or fire.
- The foundation should touch the frame at the bottom if a single bottom bar is in use. A small space should then be left at the top and sides. If a gap is left at the bottom, it is not built on by the bees, whereas the top will always be closed in.
- Where frames with wedges in the top bars and double bottom bars are in use, the foundation should hang loose between the twin bars at the bottom and be securely trapped under the wedge at the top.
- It is now common practice to buy ready-wired foundation, with the wire embedded in a diagonal pattern. Crimped vertically wired foundation is also obtainable from dealers.

THE BEE YEAR

Months in this chapter relate to northern hemisphere cool temperate climates with mild maritime influence. Adjustments need to be made for other climatic zones and, in the southern hemisphere, for the reversal of the seasons.

January

In the hive: If it is very cold, the bees will all be clustered tightly together. There will be no brood yet, but the bees are not asleep, nor are they in a winter torpor like many other animals. They are generating heat by taking in food and moving around a little.

At the edge of the cluster the temperature will seldom sink below 9°C (48°F); in the centre it will be around 25°C (77°F). In frosty weather the bees counteract the cold by short bursts of temperature-raising activity which can be heard as a loud buzzing.

It is a bad thing for the weather to fluctuate too much in January because the queen will start laying in mild weather. The worker bees then have to use their body reserves to nourish the brood, which shortens their lives.

The strength of the colony in spring depends not only on the number of bees who have survived the winter (with luck about 10,000) but also how many are lost in the early spring months.

A mouse excluder, but the pest has got in! The white patch of gnawed wood and the slightly bent metal show its method of entry

Jobs for the beekeeper: The bee-keeper should make regular visits to the apiary, just as he does in summer, to check that everything is all right.

Can mice get into the hives? Are the bees being disturbed by banging doors or waving branches of trees? Is any damp getting into the hives? Is the hive entrance blocked up with dead bees? It must be kept unblocked to allow the bees to fly out in mild weather.

And that's not all – the winter is a good time for home study and for doing practical jobs: building and repairing hives, making and wiring frames, rendering down wax for foundation, if you want to make it yourself, or perhaps for candles.

February

In the hive: In mid-February the mood in the winter cluster will be changing (this may happen later depending on the colony and where it is).

In the centre the first cells are being cleaned and receiving eggs from the queen. A small brood nest is being made.

As the temperature rises, brood activity increases. The winter bees sacrifice their fat-protein reserves to nourish the brood. On a fine day, when the temperature rises above 10°C (50°F), some bees will release themselves from the cluster and risk a flight outside, having to reorientate themselves.

The bee's rectum will be full of waste matter from the past three months. This will be excreted in

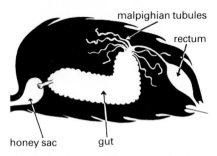

A bee's digestive tract. When an overwintering bee emerges from the hive the rectum may be noticeably enlarged, full of faeces (below)

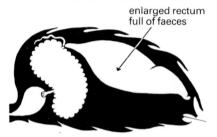

flight, leaving yellow splashes of droppings scattered around.

Jobs for the beekeeper: Check the hive entrance at regular intervals, even in bad weather. You will be able to tell a lot about the state of your bees just by a quick glance. Weak colonies can be put into nucleus boxes or much smaller hives, but only in fine weather.

If you want to move the hive within the bees' previous flying radius, you should do it before they take their first flight. There will also be work to do carried over from January.

March

In the hive: In warm areas the colony will be starting on a new cycle of life. The steadily increasing brood and newly hatched bees will consume pollen. Pollen from hazel trees, willows, alders and a multitude of spring flowers forms the basis of their development, providing there is sufficient food in the hive.

The colony consumes far more food in March than in the cold of December. In the warmth of the midday sun you can often observe the young bees making their first tentative movements outside.

The bees must have access to water close to the hive. Even in the cold and damp they will venture outside to fetch water for the brood. If the source of water is too far, they can freeze to death.

Jobs for the beekeeper: Place a source of water near to the hives, but not arranged so that the bees will defecate into it or drown in it. You could use a trough filled with moss, cork or wood wool, or a container with a dripping tap from which water drips onto a board with a small groove. The bees must always be able to find water at the watering place. To help them find it initially, mix the water with a little honey.

Without disturbing the bees, check through their rubbish. The debris on the floor of the hive can tell you a lot about the condition of the colony: how many beeways there are, how much food they have consumed, how many have died in the winter.

A sample collected from three colonies at a time should be sent to the relevant investigation department for

Handy water sources

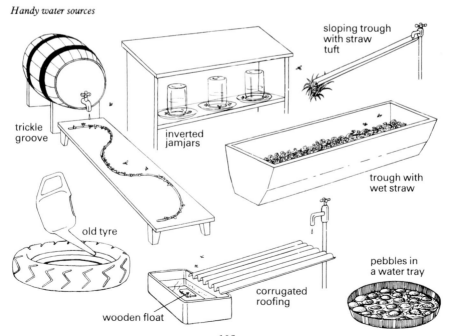

sloping trough
with straw
tuft

trickle
groove

inverted
jamjars

trough with
wet straw

old tyre

pebbles in
a water tray

corrugated
roofing

wooden float

examination for the Varroa mite. Do not inspect the colonies without good reason. Queenless colonies should be united with another colony, e.g. a new colony, and only if the temperature is above 15°C (59°F).

If the window in a rear-opening hive has misted over, the colony will be breeding and will not want to be disturbed. All they need in this case is warmth, honey, pollen and water.

If you have any pollen combs available, hang them in the vicinity of the brood nest. However, it is better to put them in earlier so that the bees have them over the winter.

Start embedding foundation into frames, or make up frames with ready-wired foundation for the coming season: about twenty per colony, including young colonies and swarms. If you have rear-opening hives, you will probably need fewer.

Take steps to ensure future honey flow when sowing and planting.

April

In the hive: With the sun rising higher in the sky and blossom time well underway, the bee colony is also increasingly active. The bees are busy foraging, building and raising new bees in time for the early honeyflow.

However, to say that the bees are at the beginning of a long foraging season is not quite true. Often the honey flow does not begin until May and ends at the beginning of July.

April shows which of the colonies will be able to survive. A colony cannot be said to have survived the winter until it has produced two generations of brood by the time the oil-seed rape is in flower.

The winter colony, with its long-lived bees, becomes a summer colony with short-lived bees. It is a good sign when this process of transformation starts late and a great mass of winter bees is available to take advantage of the early honey flow.

Jobs for the beekeeper: A fine day in April is the time to carry out the first major inspection of the colony. Often the success of early foraging will depend on this. All the colonies should be checked for the points listed on page 69 and dirt and debris removed from the floors of the hives.

If the colony is strong and large, it can be extended. Combs and drone frames, lightly sprayed with a honey solution, can be introduced into the hive – but not until the first cherry trees are in blossom, because cherry-blossom time is building time.

Hives with supers need to be enlarged once all the combs and bee-ways are covered in bees. For this purpose, add a super with three drawn-out combs in the middle, a drone frame one in from the end and frames of foundation in between.

Rear-opening hives can be extended by gradually adding frames of foundation and empty frames. The frames of foundation are hung by the brood nest, taking care not to disturb anything in this 'holy of holies'.

Any colonies which are unsettled or queenless should be united with others. For queenless colonies this is easy: the bees are shaken onto a cloth near the hives and then beg an entrance into other colonies.

Drone larvae exposed when the queen excluder is pulled off

Weak colonies will have a bad queen, and it is advisable to find her and kill her. The bees can then be treated as above. Any brood comb or comb filled with food can be hung in other hives.

In a hive with supers a complete colony can be united with another quite easily. The queenless colony is simply put on top of another colony in its original hive. A newspaper soaked in honey, with small holes in it, is placed between the two. This allows the colonies, with their different odours, to get to know each other slowly. Gradually they bite through the paper.

As a rule of thumb, it is important to remember that strong colonies grow stronger and weak colonies weaker. So there is little point in uniting two weak colonies.

If the weather is bad and the honey flow late, it may be necessary to feed the bees. This is a time when both bees and brood need lots of food. They must get it before they have an acute shortage, so they may need to be fed as early as March.

From April onwards you will need to check the quality of the combs. With the rising temperature any wax moths that have survived the winter will be active again.

May

In the hive: Brood rearing is in full swing. The bees are busy collecting

A swarm resting on a horse chestnut branch

pollen and nectar. To take advantage of an early honey flow, the brood chamber should be full of bees.

Jobs for the beekeeper: The growing colony needs room, so introduce frames of foundation as they are needed. In hives with supers these frames can be hung near the brood nest where they will be drawn out sooner. Once a frame has been drawn out, you can swap it for one on the outside. Every beekeeper develops his own working pattern for this, but the golden beekeeping rule is never to fiddle with the brood nest.

If, despite a good honey flow, a colony slackens off its building work, this is an indication that swarming may be imminent. In this case, put in empty combs so that the queen can lay immediately.

As soon as there are bees on all the combs, it is time to consider whether the brood nest needs expansion, as well as the need to add one or more honey supers over the queen excluder. The brood nest can be extended by moving two or three frames of capped brood into a second deep box just above the queen excluder, making sure the queen stays below.

The capped brood frames are then replaced with foundation or empty drawn combs. Alternatively, the queen can be given access to a second deep or shallow box under the queen excluder, so that she can work over twenty or more frames.

Honey supers must be added ahead of the bees to give them house and storage room. Some beekeepers dispense with queen excluders. They take out frames of honey as the bees fill them, for extraction, and return them to the hive, or simply pile on more honey supers if all the beeways are full of bees.

The honey flow from oil-seed rape can bring on the swarming impulse. If you want to stop the bees swarming, you may need to fit a special dividing board, as described on page 74. As it is not always possible to stop swarming, you will have to be ready to cope with any swarms that do occur. It is also important not to forget about building up young colonies.

June

In the hive: The summer solstice on 21 June marks the climax and turning point for the bees. The swarming impulse is subsiding now. Given fine weather and a good honey flow, the bees are fully occupied preparing honey. The colony has now grown and matured and is laying down supplies for the winter.

However, in southern Britain there is often a marked dearth of forage for the bees in early June called the 'June Gap', between the spring and the main summer flowers.

Jobs for the beekeeper: If the bees have found oil-seed rape in May it is now harvest time. Once the rape fields turn from yellow to green the rape honey must be extracted at once.

Only ripe honey should be harvested and the beekeeper should leave the bees enough floral honey which is essential to keep them healthy. Rape honey must not be left in the combs as it sets hard, so that even the bees cannot make use of it.

Once the rape honey flow is over many beekeepers move their bees to bean fields and other areas of better honey flow. Woodlands can provide a really good source for foraging. The beekeeper might also consider moving his or her hives to areas with crops of maple, robinia and sweet chestnut, and later pine and spruce trees.

In times where there is no honey flow, robber bees can be a problem. Robbing, however, is something that the beekeeper can prevent, and if it does happen, it is probably because of his own stupidity!

If you have to work with the lid off the hive, do so as quickly as possible; do not leave combs out uncovered, nor splatter honey and food around. If alien bees find even the slightest drop of honey, they sense a food supply and in a trice hundreds will be

Hives at the edge of an oil-seed rape field

buzzing around, obstinately looking for food.

They can easily fly into the open hive and rob it. The bees in the hive will naturally defend themselves, which may leave many bees dead, often including the queen. Weak colonies are particuarly at risk.

If bees start to rob a hive, stop work on any tasks on the hive immediately and wait for a couple of hours. You may wish to try chasing off the bees with a spray.

If a colony is completely robbed, an event which can often happen to weak colonies in spring, you should clean out the hive completely and leave it standing empty, otherwise the robber bees may home in on neighbouring hives and kill off other colonies.

July

In the hive: In Britain early July is the peak of the summer flow when the bees store most of their surplus ready for the winter. Brood rearing has already passed its peak, so there are more foraging bees and fewer young mouths to eat up what they gather. Some colonies stop brood rearing altogether for a time.

As part of a prudent long-term strategy, some of the emerging bees will be 'unemployed'. They will be laying down reserves of fat and protein in order to become winter bees. This process is helped by the spasmodic honey flow in later July and August.

If nearby woodland areas are producing a honey flow, however, the story is different. Then, the bees will

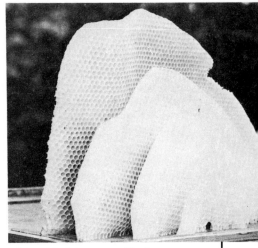
Wild comb built in an empty hive box

devote all their energy to foraging and will probably not survive the end of the honey flow.

Bee virus paralysis can often occur now, particularly if there is a dearth of pollen. A prudent beekeeper will already have prepared for this eventuality by making young colonies.

At the very end of July the first drones are being driven off.

Jobs for the beekeeper: In some areas and in some very wet or dry seasons the nectar flow may dry up altogether, in which case the unfortunate beekeeper must take care that densely crowded colonies don't starve to death. Every colony should have at least two full combs of honey in store.

August

In the hive: The main honeyflow has now ceased, and many more drones will be given notice to quit.

Some later flowering plants will still give occasional flows, but the danger of robbing by unemployed bees continues to increase. Brood rearing is at a low point.

Jobs for the beekeeper: He still continues to watch for robbing, but towards the middle of the month will begin harvesting the main crop of honey.

When all the harvesting is done and the honey supers are removed, start to arrange the brood nest and combs for the coming winter. Take out all frames of foundation which have been only half completed, and old dark combs and replace them, if possible, wth light-coloured combs in a clean and tidy state, which have been used for brood before.

Colonies should winter on a minimum of eight combs. Nucleus boxes should not go into winter on fewer than five combs. If there is likely to be no more honey flow, winter feeding can begin. Young colonies in particular will benefit from this.

If the beekeeper can move the bees to heather, this should be done at the beginning of the month to make maximum use of this resource.

September

In the hive: The bees are preparing the hive for winter. Food is stored around the brood nest, gradually hemming it in. The last of the drones are driven off. Where this does not happen, it is a sign the colony wants a new queen.

On no account should you remove queen cells, particularly if they are capped. Supersedure is the great virtue of bees: it is often a mistake for a beekeeper to think he is doing the bees a favour by introducing a new, expensive, thoroughbred queen.

Jobs for the beekeeper: Feeding should now be completed. All the colonies should have their winter supplies laid down by mid-September. For colonies with food in two hive boxes, it is important that all the brood frames are in the bottom of the hive, as the bees eat their way upwards.

Now is the time to carry out a Varroa examination if this has not been done before.

October

In the hive: As in all months, the weather will be the determining factor as far as the bees are concerned. If it is still fine and there is a late honey flow, the bees may produce one more lot of brood. The pollen flow is important for the colony's development next year. All the pollen supplies should be left in the colonies, although they can be redistributed among them.

If the weather is cold, the bees will fly out only when they have to. They will begin to huddle in their winter cluster, coming apart if there is a period of finer weather.

Jobs for the beekeeper: The bees need feeding up. If by chance the area yields an extraordinary late honey flow in September (heather for example) there will be a chance to extract some late honey and to feed them up with syrup as quickly as possible. It is often a good idea to give such late colonies capped combs of food from other colonies.

November, December

In the hive: In this season the bees turn in on themselves. During the day they might risk a last sortie. The first frosts at night and temperatures of around 8°C (46°F) make them draw together in their winter cluster. The last of the brood will have emerged in October.

If the bees are dry, they can survive the coldest weather without damage, but there should be at least 10,000 bees in the cluster. The bees are constantly changing places, allowing the bees on the outside through into the warmth inside the cluster.

Providing it is warm enough, the bees will still make occasional flights to defecate. This is very important, because if they cannot relieve themselves for a long time, there is a risk of amoeba and nosema diseases.

If it snows, there is no cause for concern, even if the hive is completely covered. The bees will not suffocate, in fact the snow helps to keep them comparatively warm. The only danger is if the hive entrance ices over completely, but this seldom happens.

Even carbon dioxide levels of 6 per cent will not harm the bees (in normal air the level is 0.03-0.06 per cent).

A cold winter is good for them, but changeable weather is dangerous, as they might start producing brood too early and have a setback if there is another cold spell.

Jobs for the beekeeper: If the weather is good enough, Varroa treatment should be carried out. Protect the bees from cold and damp by putting a layer of insulation on top of the hive.

The hive entrance should be made smaller to deter robber bees. It is a good idea to do this when starting to feed the bees. In winter shrews and mice are often unwelcome guests in a hive. A 7 mm (¼ in) mesh can be placed over the hive entrance to ensure none can get in to feed off the bees.

To keep birds away, it is best to feed them at some distance from the apiary.

All good surplus combs should be stored in cupboards or empty hives with a small amount of acetic acid left to evaporate.

Brood combs ruined by overwintering mice

HONEY PLANTS

Apart from looking after his bees properly, a beekeeper's main concern is to ensure the bees have access to an adequate honey flow from a wide variety of plants.

Any enrichment of plant life will have a direct effect on the bees; on the other hand, so will any deterioration. Even a small crop can make a vital difference to the bees' health. You do not necessarily need your own garden: wasteland, quarries and hedgerows can all be made use of.

If you manage to optimize the source of food for your bees and the weather is all right, you will almost certainly have healthy bees and a good honeycrop. One of the most enjoyable aspects of beekeeping is finding out whether and when different types of plants thrive and which ones are attractive to bees.

The bees will have particular plants they love to visit and others they don't like at all. Different plants produce different amounts of nectar and pollen. Some provide the bee mainly with sweet nectar, others give only pollen. The various food soures can be divided roughly according to:

1 Time of flow: early flow, summer flow and late flow.
2 Type of flow: pollen, nectar or honeydew.

Early, Summer and Late Flows

The **early flow** stretches from April to June. Crocuses, hazelnuts and willows provide pollen; willows, fruit trees, dandelions and oil-seed rape give nectar as well.

Because of modern farming methods the **summer flow** is scarcely adequate for bees these days. Weeds and wild flowers, like cornflowers and wild mustard, used to provide an abundant supply of nectar and pollen. Spruces, Norway maples, horse chestnuts, robinia, lime trees, alders, brambles, sainfoin, white clover and wild sage often provide welcome sources of forage.

The **late flow** can provide a rich source of pollen from flowering plants such as yellow mustard, oil-seed rape or phacelia (California bluebell). Silver firs and spruces also often have a late honey flow, as does ling.

Pollen

Bees visit the plants at specific times of the day to collect pollen once the stamens on the flowers have opened. The pollen is mixed with saliva and honey and carried back to the colony in the bee's pollen baskets.

It is always a lovely sight to see your bees returning with full pollen baskets. The range of colours of the pollen in the baskets can be astonishingly wide. The bees need the pollen to raise brood and young bees and it has been estimated that a single colony requires about 30 kg (66 lb) each year.

Pollen should be available to the bees from early spring to late autumn.

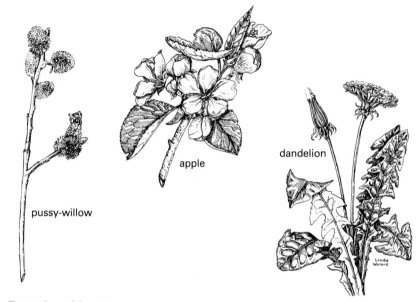

pussy-willow

apple

dandelion

Forage plants of the spring

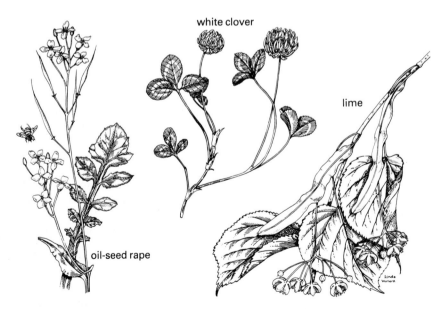

white clover

oil-seed rape

lime

Forage plants of the summer

114

Despite the rate of growth of the colony, the bees will find themselves with a surplus in May/June, which they lay down as stores.

They tamp the pollen down in the cells so firmly that it cannot be dislodged by the honey extractor. It then undergoes lactic fermentation and is sealed over with honey, which preserves it.

There is no ideal substitute for pollen, although in an emergency you may feed your bees with one of the proprietary pollen substitutes on the market.

The quality of pollen varies from plant to plant. Pollen from willows, fruit trees, clover, oil-seed rape and crocuses is considered good; from alders, birch trees and grasses less good; and from conifers worthless. The pollen from horse chestnuts and some ling and buttercup varieties can be toxic, although there is little danger from this toxic effect providing the bees have a supply of pollen from many different sources.

The main pollen-yielding plant families are:

Rosaceae: pomaceous fruits,
   berries.
Crucifers: oil-seed rape,
   mustard, brassicas.
Compositae: dandelions,
   sunflowers, cornflowers,
   thistles, asters.

Nectar

Many plants secrete a sugary liquid, called nectar, from their nectaries. Nectar is a watery solution containing an average of 20 per cent sugar, generally raw sugar. The water con-tent can vary between 30 and 90 per cent. It also contains minerals, amino acids and vitamins.

The quantity and composition of the nectar varies according to the specific type of plant. There are still many unanswered questions about its formation. What is clear, however, is that warm and damp air encourages nectar production and dry winds have the reverse effect.

A colony needs about 40 kg (88 lb) of honey each year, which corresponds to about 225 kg (496 lb) of nectar. The bee uses its proboscis to suck the sweet liquid out of the flower and into its honey stomach.

On its way back to the hive the bee sets to work immediately to add an enzyme from its own body to the freshly collected nectar. This enzyme has the effect of splitting the raw sugar into glucose and fructose.

The nectar/honey mixture, still undergoing this reaction, is passed from the foraging bees to the house bees and stored in the cells. The honey is moved around several times. Water has to be evaporated off the mixture and further enzymes and bacteria-preventing substances added. Gradually a chemical transformation occurs, changing the sweet watery nectar into golden honey.

A cell with ripe honey is sealed with a thin, light-coloured layer of wax to make it airtight. This is important because open honey is hygroscopic, i.e. it readily absorbs and retains water. Preparation and storage of food in this way is peculiar to bees.

In order to ensure the bees have a surplus of honey, all the conditions must be right:

Table of Honey Plants

The following table lists the most important honey plants, their flowering periods and their nectar/pollen yield.

Flowering time: 1 – 12 = number of month. Northern hemisphere: 1 = January, 12 = December. Southern hemisphere: 1 = July, 12 = June
Yield: 1 = low, 4 = very good
★ Honeydew produced

Early Honey Plants				
Name	**Flowering time**	**Nectar**	**Pollen**	**Honeydew**
Anemone	3–9		2	
Alder	1–3–5		3	★
Ash	4–5		2	★
Beech	4–5		3	★
Berry bushes				
Blackcurrants	4–5	2	1	
Brambles	6–9	3	3	
Gooseberries	4–5	3	2	
Raspberries	5–7	4	3	
Birch	4–5		2	★
Bird cherry	5–6	2	2	
Birthwort	3–5	2	3	
Blackthorn	4–5	2	3	
Borage	5–7	3	2	
Buckthorn	5–6	2	1	
Cornelian cherry	2–4	3	2	
Daffodil	2–4	2	2	
Dandelion	4–5	3	4	
Dogwood	5–6	2	1	
Elm	3–4		3	★
Forget-me-not	3–6	2		
Fruit trees				
Pomaceous fruit				
Apple	4–5	4	4	
Pear	4–5	2	3	

Early Honey Plants

Name	Flowering time	Nectar	Pollen	Honeydew
Stone fruit				
Almond	3–4	3	3	
Apricot	4	2	2	
Cherry, sweet	4	4	4	
bitter	4	4	4	
Peach	3–4	2	2	
Plum	4–5	2	2	★
Goutweed	6–7	2	1	
Hawthorn	5–6	2	2	
Hazel	2–3	3	2	★
Hellebore	12–4	2	3	
Horse chestnut	5–6	3	2	★
Lungwort	3–5	1	1	
Maple				
Common maple	4–5	2	1	★
Mountain maple	5	4	2	★
Norwegian maple	4–5	3	2	★
Oak	4–6.		3	★
Oil-seed rape	4–5	4	4	
Poplar	3–4		3	★
Robinia	6	4	2	
Rowan	5	2	2	
Sage	6–8	3	1	
Sedge	4–6		1	
Snowdrop	2–3	2	2	
Wild chervil	4–6	2	2	
Wild garlic	5	2	1	
Willow	3–5	4	4	

Summer Honey Plants

Name	Flowering time	Nectar	Pollen	Honeydew
Brassicas	5–7	3	2	
Buttercup	3–9	1	1	
Carrot	6–9	2	1	
Centaury	6–8	3	2	
Charlock	6–7	3	2	
Clematis	5–9		2	
Clover				
Alsike	5–9	4	3	
Crimson clover	5–7	3	3	
Red clover	6–9	3	3	
White clover	5–10	4	3	
Common spruce	5–6		2	★
Corn	6–9		4	
Cornflower	5–10	3	3	
Cow parsnip	6–9	3	1	
Creeping thistle	7–8	3	2	
Crowfoot	5–9	3	1	
Fir tree (silver)	6		2	★
Germander	7–8	3	1	
Grasses, cereals	5–7		2	
Hollyhock	7–9	3	1	
Horse bean	5–7	2	2	★
Knotgrasses	5–10	1	1	
Larch	3–4		1	★
Lime				
Silver	7	3	1	★
Summer	6	3	1	★
Winter	7	3	1	★
Lucerne	6–9	3	1	
Meadow rue	5–8		2	
Meadowsweet	6–7		3	
Mustard	6–7	2	2	
Parsnip	7–8	1	1	
Pine tree	5–6		2	★
Plantain	5–10		3	
Poppy	5–8		4	
Sainfoin	5–7	4	4	
Sorrel	5–8		2	

Name	Flowering time	Nectar	Pollen	Honeydew
Sunflower	7 – 10	3	3	
Sweet clover	6 – 9	4	3	
Vetchling	6 – 9	2	1	

Late Flowering Plants

Name	Flowering time	Nectar	Pollen	Honeydew
Autumn dandelion	7 – 10	2	1	
Buckwheat	6 – 9	4	2	
Common heather	6 – 10	4	1	
Corn	6 – 9		4	
Erica	6 – 9	3	1	
Goldenrod	7 – 9	3	2	
Hibiscus	7 – 8	3	1	
Ivy	9 – 10	3	1	
Mint	7 – 9	2		
Plantain	5 – 10		3	
Red clover	6 – 9	3	3	
Virginia creeper	7 – 9	1	2	
Willow herb	6 – 9	3	2	

Urban Plants

Name	Flowering time	Nectar	Pollen	Honeydew
Ailanthus	7	3	2	
Cordon	8	3	2	
Flowering ash	5 – 6	1	3	
Horse chestnut	5 – 6	3	2	
Privet	6 – 7	2	1	
Robinia	6	4	2	
Sweet chestnut	6 – 7	4	3	★
Virginia creeper	7 – 7	1	2	

Crops

Name	Flowering time	Nectar	Pollen	Honeydew
Asparagus	4–5	3	3	
Brassicas	5–7	3	2	
Buckwheat	6–9	4	2	
Carrot	6–9	2	1	
Corn	6–9		4	
Dragon's head	5–7	3		
Fennel	6–7	3	2	
Flax	6–7	1	1	
Hedge mustard	5–7	3	2	
Hedysarum	5–8	3	2	
Hemp	6–8		2	
Horse bean	5–7	2	2	★
Leeks	5–7	3	1	
Lucerne	6–9	3	1	
Lupin	6–9		2	
Melon	6–8	2	1	
Mustard	6–7	2	2	
Oil-seed rape	4–5	4	4	
Parsnip	7–8	1	1	
Pea	5–6	1	1	
Poppy	5–8		4	
Sanfoin	5–7	4	4	
Serradella	6–7	3	1	
Squash	6–8	3	2	
Sunflower	7–10	3	3	
Sweet chestnut	6–7	4	3	★
Turnip	4–5	4	2	
Vetch	5–7	1	1	

- fine foraging weather;
- a large crop of plants close to the hive, producing nectar or honeydew;
- sufficient numbers of foraging bees.

The major nectar-yielding plant families are:

Rosacaeae: pomaceous fruits, stone fruits, berries.

Labiates: wild sage, thyme, marjoram, lavendar, dead nettles.

Compositae: dandelions, cornflowers, sunflowers, goldenrod.

Papilionaceous plants: white clover, alfalfa, sainfoin.

Honeydew

Honeydew is also a sweet, watery solution, but it contains protein as well. It is secreted not by the plants themselves, but by types of aphids and coccids, or scale insects which live on various kinds of woody and herbaceous plants. They pierce the phloem of the plants and suck the sap out of the leaves, shoots and buds.

Honeydew contains 16-20 per cent sugar, amino acids and many minerals. The aphids need only 5-10 per cent of the carbohydrates for their own consumption and only half the amino acids (protein components); the rest they secrete, after converting the plant sugar.

This means that honeydew which has come from the same plant, but has been processed by different plant-sucking creatures, can have a completely different composition. This explains why bees may shun honeydew one year but collect it quite happily the next: it has simply been produced by a different bug.

Aphids generally spurt out the sweet juice, leaving sticky, shiny droplets on the leaves. In the case of scale insects, the honeydew works loose and rolls off.

The honey crop depends on how well the insects are doing; this in turn depends on the weather and the condition of the trees. Hot, dry summer days without a morning dew, a sudden drop in temperature, downpours and hail are bad and cause the honeydew to dry up. A high water table, humidity and a river in the vicinity are all beneficial.

The importance of these woodland sources has increased in recent years with the decline in the honey flow from flowers, and the increasing popularity of wild honey. Many German beekeepers now use the floral honey flow to build up the colonies in order to collect the woodland honey sources.

There are about a hundred different honeydew yielders in total; every tree type has one or more and the beekeeper should get to know the most important of them. In Europe the major providers of honeydew are certain species of scale insect and aphids of the Lachninae sub-family.

In contrast to aphids, scale insects produce only one generation of offspring per year. In winter the female larvae can be found hidden under the whorls, and the male larvae as scales on the needles of the host tree. In spring the larvae moult and become adult insects.

The females remain where they are, little 'buttons' in the whorls of the trees. They live from April/May to July. They generally secrete the most honeydew in mid May. Honeydew from scale insects is generally at its most abundant at the time the elderberry is in flower.

In contrast to the supply from aphids, the flow is generally quite short, often lasting only eight to fourteen days. The bees should therefore be moved to the woodland well before the honeydew is expected.

The male insects are small and can move about. After mating, about 200-1,000 larvae start to develop inside the swelling female. These hatch between July and September after the death of the mother, and overwinter in the whorls of the needles.

One major difference between aphids and scale insects is that scale insects start reproducing the year before, whereas aphids do so shortly before producing honeydew.

The most common scale insects are two types of spruce-bud scale:

Physokermes picaeae, which prefers old wood and thin shaded branches. It is one of the most reliable and abundant producers of honeydew in the pine forest. The reddish brown insect is difficult to spot because it looks like a bud and sits on the youngest whorls. Some species of scale insect can also be found on other trees, like thuja and oak.

Physokermes hemicryphus, which prefers more vigorous branches. The honeydew it produces is collected mainly by ants.

Aphids overwinter as eggs. In April the egg develops into a wingless female which produces live female offspring in May.

In June they may produce a second generation of females and, under certain circumstances, a third. When the food supply, i.e. the nitrogen content in the sap, decreases, the second generation may be winged. These winged insects can then fly to other areas and produce another unexpected supply of honeydew. The last generation, in October, produces winged males which mate with wingless females. Eggs are laid up to the beginning of November.

The reproductive cycle doesn't always follow the same pattern; in the course of the year there will be different rhythms. The insects also have their own enemies: animals like the ladybird and the larvae of lacewings and ichneumon flies.

Ants, particularly the red ant, look after the aphids. Honeydew secretion increases considerably in the vicinity of ants' nests even though the ants are consuming it as well. By touching the aphids with their feelers, the ants stimulate them to produce more honeydew. Ants also even move some types of aphid to a good source of food, attending them and protecting them from enemies. It is therefore worth the beekeeper's while to protect the ants in his turn.

The most important aphids for honeydew are:
- The reddish brown *Cinara pilicornis* which produces a first generation of females early in the year and yields the first honeydew. If there are ten to twenty colonies per tree by the end of May, the tree is likely to produce a good crop of honey.
- The black *Cinara piceae*. 4-5 mm

rosebay
willow herb

ling
(Calluna vulgaris)

ivy

Linda
Waters

blackberry

Forage plants of late summer and autumn

(⅕ in) in size, it generally sits in groups between the needles of the spruce, sucking at the bark on thin branches and on the trunk. It produces a good crop, generally at the beginning of June, but occasionally not until July. As several generations of daughter aphids occur in a single year, honeydew may be produced given the right weather, up to September.

- The green *Cinara pectinatae* is 5 mm (⅕ in) long and has two white stripes on its back. It lives not in colonies, but individually on needles, from which it sucks juices. It doesn't like the light and spends the winter on the underside of the needles.

The honeydew crop can stretch from May until September. There will generally be a bumper harvest every four to six years. This green aphid is the most important honeydew yielder for fir trees. There is also a black variety.

BEE DISEASES, PESTS AND PREDATORS

Intensive methods of animal husbandry have increased the incidence of diseases and pests in animals. Thoroughbred honeybees are particularly at risk. Unfavourable conditions can cause parasites and germs, which may be lying dormant in the colony, to multiply and spread.

Providing a beekeeper keeps his bees properly and the honey flow is good, he has nothing to fear. He must know, however, the possible dangers facing bee colonies and which diseases have to be reported.

Brood Diseases

American Foulbrood

American foulbrood is a notifiable disease: it must be reported. It is caused by *Bacillus larvae*, a rod-shaped bacterium which divides up into innumerable spores. In its inactive spore form it can be taken with food into the gut of larvae. Here the spores develop into the harmful bacteria which kill the larvae. The *B. larvae* then reverts to its spore form and can survive for many years.

Often, by the time the symptoms have been spotted, it is too late to save the colony. The cappings over brood cells may be sunken and perforated.

To check for the disease a matchstick can be pushed into the dead larvae under these cappings. If diseased, the larva will be light brown in colour and form a 1 cm long (⅓ in) ropey, foul-smelling mass which will cling to the matchstick as it is withdrawn. A dark-brown, hard scale, which can be made up of thousands of millions of individual foulbrood spores, will be left sticking firmly to the lower side of the cell.

The infection can be sparked off by spores carried by drifting bees, from derelict apiaries or waste dumps, or from within the derelict hive itself. Foulbrood can also be brought in through feeding with foreign honey, so it is advisable to use only your own honey for feeding purposes.

European Foulbrood

European foulbrood, which has a number of causative agents including *Melissococcus pluton* (formerly *Streptococcus pluton*), generally occurs in late summer. It causes the collapse and decay of the larva.

Infection occurs while the larva is at the round grub stage. In an infected brood, the lower angle of the cell will contain a dry scale. If the matchstick test is done, no ropey threads will stick to it. The scale can be loosened and the smell is noticeable.

The disease is less serious than American foulbrood but is notifiable.

Chalk Brood

Chalk brood is the most common

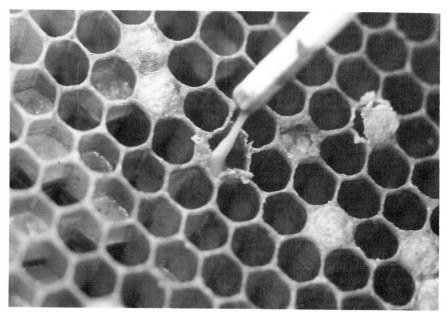

The 'rope test' for American foulbrood

disease, but causes the least extensive damage. It can often be traced back to chilling, for example when there is a lot of brood but, because of a sudden cold spell, there are not enough bees to keep them warm.

The micro-organism which causes the disease is a fungus called *Ascosphaera apis*. It is present in all colonies, but is harmful to the brood only if there is too much dampness in the hive. The fungus is carried in spore form, via food, into the gut of the larva, causing the latter to dry out. It can be confused with old mouldy pollen, which it resembles closely.

The black-and-white mummified remains of the larva are usually left in place but occasionally may be carried by the bees to the hive entrance. Affected brood combs rattle when shaken.

If the disease occurs, the beekeeper should leave the healthy brood to hatch in the honey chamber and feed the bees with a ½ litre (1 pt) of warm honey solution in the evenings to encourage them to clean. It may be necessary for the beekeeper to make the colony smaller and rethink his beekeeping practices.

Sacbrood

Sacbrood is a rare disease. Generally speaking it is relatively harmless, but given a poor honey flow or bad weather, it can become epidemic. The disease is caused by a virus, *Morator aetatulae*, which is transferred to the young larvae in the feeding process. The infection usually strikes two-day-old larvae, enveloping them in a bag of fluid. The remains can be removed from the cell with tweezers. In the course of time the fluid dries up and the larva's body collapses, leaving a brown scale on the cell floor. If there are just a few cases, the combs can be hung in the honey chamber and burned or melted down after the bees have hatched. Otherwise the colony should be kept warm and fed a honey solution to stimulate their cleaning instinct. If there is a severe outbreak in a strong colony, the best idea is to form an artificial swarm; badly affected colonies should be given a sulphur treatment.

Diseases Affecting Brood and Adult Bees

Stonebrood is also caused by a fungus, *Aspergillus flavus*, and the initial symptoms are similar to those of chalk brood. It is, however, mercifully rare.

The disease turns the larvae into hard mummies, firmly attached to the cell, making it impossible for the bees to remove them. The hardened mummy can form greenish yellow spores, resembling pollen spores, around the head end of the dead larva. Healthy bees can cover dead larvae with propolis.

The fungus can also attack adult bees, gluing up their abdomens. This makes them unable to fly and they can be found scrambling around in front of the hive.

Although the disease is rare, it needs to be watched carefully because it is the only bee disease which can also be transmitted to humans, causing protracted inflammation. If the disease is suspected, the relevant authorities should be notified.

Varroa

Varroa (notifiable) is a parasitic disease which is currently the major problem in beekeeping throughout the world.

The parasite is the *Varroa jacobsoni*, a spider mite. The brownish mite is visible to the naked eye, the female being 1.2 mm long and the male slightly shorter.

It is oval in shape, the back being slightly arched and covered in thick hair. The four pairs of legs have adhesive pads. The female mites generally attach themselves between the segments of the bee's abdomen, where they remain through the winter.

All reproduction occurs in the bee brood cells and the development cycle is similar to that of the bee brood. The impregnated female looks for a brood cell about to be capped, preferably a drone cell. After the cell has been capped, she starts to lay eggs.

A female takes seven to nine days to develop, a male five to six days. Mating takes place in the capped cell, the male dying afterwards. The impregnated females leave the cells with the emerging young bees.

The mites suck the haemolymph (blood) of pupae and adult bees. As a new infestation cannot be easily detected, vigilance is essential. Signs to watch out for are a patchy brood nest, bees emerging malformed,

Varroa mite, viewed from top and underside

pupae and crippled bees being ejected from the hive, generally restless behaviour and poor flying ability or a weakening of the colony.

Mites on the bees can also be detected visually. Diagnosis, once the infestation is advanced, is simple: in July/August floor inserts can be put into the hive and examined after a few days for dead mites.

If the disease is suspected, debris from the hive must be sent to the appropriate authorities for examination. Treatment must be carried out in accordance with the relevant legal provisions. At present new biological methods are being researched.

Varroa is no reason to give up beekeeping, but it is certainly a challenge to us all! It has recently been found in the USA and it is likely to make its way to Britain. Vigilance is essential.

Diseases Affecting Adult Bees

Acarine Disease

Acarapis woodi is a parasitic spider mite not more than 0.15 mm long. It lives and multiplies in the bee's breathing tubes or tracheae, primarily in the tubes of the prothorax. It cannot exist for more than a few hours without the bee.

The infection takes hold in the first days of the bee's life, while the brushes on the spiracles are still soft and pliable. The mite can be passed on via young bees, equipment and comb. It sucks its nourishment from the tissue in the bee's air tubes,

Inspection for acarine disease. The beekeeper performs a 'post-mortem' if the disease is suspected, opening the thorax to look for the signs

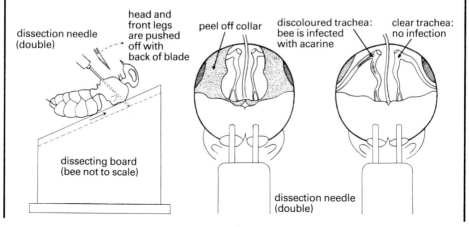

leading to a decline in the bee's performance.

Bees may be seen crawling around the hive entrance. The symptoms generally manifest themselves after the winter rest period, but a firm diagnosis entails microscopic examination of the tracheal system.

Nosema Disease

Nosema disease generally occurs in the spring and can do extensive damage. It is caused by a one-cell parasite which gets into the midgut of the bee along with food, and multiplies there, causing digestive disorders.

Nurse bees are still able to feed the larvae, but in doing so they use up all their body reserves, are unable to fly and die prematurely because they cannot build up any more protein reserves.

Nosema spores from a bee's gut

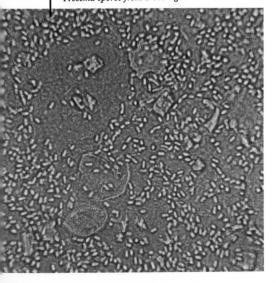

The spores are excreted in the faeces which the beekeeper will notice on the walls of the hive as yellowish brown, watery splashes.

Diagnosis can be confirmed by removing the gut from the abdomen; if the midgut is milky rather than flesh-coloured, and smells unpleasant, the bee is likely to be infected with nosema. A smear of infected fluid from the gut shows up under 400 × magnification as having thousands of rice-grain-shaped spores.

The disease can strike anywhere and does most harm to weakened colonies. Ways to help prevent the disease are:

- find a good location with an abundant year-round source of food;
- look after the bees well in late summer and start feeding them in good time;
- allow only good, strong colonies to overwinter;
- don't push the bees to produce too much brood in spring;
- cover water supplies;
- use of a specific antibiotic, Fumidil 'B'.

Dysentery

During the winter digestive problems may cause diarrhoea, which is why dysentery is often confused with nosema. The bees empty their overworked colons in the hive. Dysentery is not infectious like nosema.

The disease can be prevented by leaving a plentiful supply of floral honey in the hive, removing the honeydew honey and feeding the bees in good time (camomile tea may help). Avoid disturbing them in

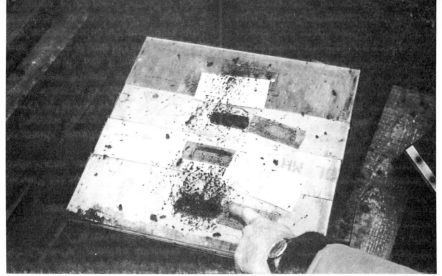

Signs of dysentery on the crown board of a hive

winter, as any disturbance will make them increase their food intake.

Bee Virus Paralysis

Bee virus paralysis often strikes if there has been a good honeydew crop from the woodland areas. Honeydew rich in minerals, a lack of pollen and the presence of bacteria can all combine to disturb the bees' metabolism. They lose their vitality and shed their hair, which makes them look black.

For treatment the bees need to be removed from the woodland and fed floral honey which often provides an instant cure. Alternatively, requeening often cures it, as the new queen may impart a greater genetic resistance to her progeny.

Pests

The **bee louse** (*Braula coeca*) is a 1.5 mm long, oval-shaped wingless fly with three pairs of legs. The lice generally attach themselves to the bee's thorax and migrate from there to the mouthparts, taking food from the bee's proboscis.

Bee lice prefer queens; if a queen is badly infested, she may stop laying eggs. The lice can be detected when examining for Varroa.

The **bee beetle or bee wolf**, found in continental Europe, is a digger wasp with a black and yellow patterned body. The female is 12-16 mm (⅔ in) and the male 10-11 mm (½ in). Only the female, which catches bees, is dangerous. She stings, paralyses

Bee louse – Braula coeca

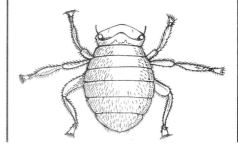

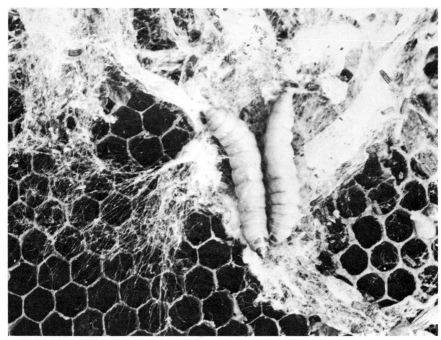

Wax moth larvae and cocoons

and sucks at her victims, or feeds them to her larvae. In the event of a severe attack, it is probably advisable to move the hives and let the bee beetles starve.

Wasps can often be very troublesome to both the beekeeper and the bees. They can be caught in a bottle, with a narrow neck, filled one-third full with sugared water and a little beer or cider.

The **greater and lesser wax moth** are among the few living creatures that can digest wax. Both species are an inconspicuous grey-brown colour. The female moths penetrate the bee colony, or places where the combs are stored, and lay clusters of eggs on the combs.

The eggs develop into white larvae about 2-3 cm (1 in) in size. These eat the wax and leave a channel through the comb with a fine grey thread and black faecal deposits. The larva pupates in a white papery cocoon, to emerge after about forty-five days.

At temperatures below 9°C (48°F) the moth's development comes to a standstill, but eggs and larvae can withstand temperatures below 0°C (32°F). A sharp frost, or twenty-four hours in a deep freeze, kills the eggs and larvae. Under certain circumstances the whole process of development, from the laying of the egg to the hatching of the moth, can take three months, so that eggs laid in autumn can overwinter.

Strong, healthy bee colonies can cope with wax moths, providing there

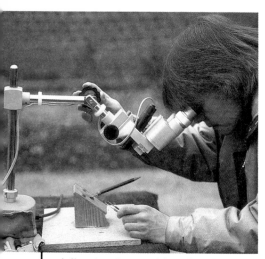

A dissecting microscope (above) is used for diagnosing acarine. A higher-powered microscope with slides (right) is used to detect nosema

are bees occupying all the combs. Combs should be stored in moth-tight cabinets.

For treatment, paradichlorobenzine is available commercially, but leaves a residue.

The common **garden ant** can be a great nuisance when colonies of them invade a hive and eat the honey. In warm climates (not the UK!), to stop them getting in, bee-houses should have an ant-proof base or pedestal containing a channel filled with water or diesel oil which prevents the ants crawling any further.

Shrews, particularly dwarf shrews which are in danger of extinction, are useful insect eaters in summer. In the winter, however, they can slip into hives and eat bees. To stop them getting in, the hive entrance should be made smaller or a mesh (approximately 7 mm [¼ in] diameter) placed over it.

Other Inhabitants of the Hive

There are a whole series of small animals which creep into the hive without causing undue damage: earwigs, silverfish, beetles, pseudo-scorpions and various mites. They generally feed off the rubbish in the hive; the pseudoscorpions even eat up the eggs of wax moths and bee lice. But some types of mites can eat pollen supplies too. Cleanliness and the regular removal of rubbish and dead bees keep these creatures under control.

Dangers from Agrochemicals

Insects cause a lot of damage to our crops, but the insecticides used to control them also do a lot of harm to other, useful insects.

Contact poisons attack the nerves, muscles and circulation. The bees suffer painful convulsions. Symptoms of spasms and cramp are a clear indication of poisoning. It is mostly field bees which are affected, but house bees can also suffer if the field bees bring the chemicals back on their coats.

If the poisoned source of food is nearby, heavy losses can occur and the hive entrance may become blocked with dead bees. Ingested poisons are often brought into the hive with pollen. They generally cause chronic disorders which are difficult to spot.

Deaths from poison may stretch over several weeks, but young bees will probably die very quickly. If this happens in June, the colony will probably be so weakened that it will not recover in time for winter.

Beekeepers should report spray damage to their local beekeepers' association as soon as they notice it.

Treatment for Cases of Poisoning

If you suspect your bees are being poisoned by agrochemicals, you should try to find out who is responsible. It may be possible to call in an advisor from your agricultural or local authority to collect a sample of, say, a thousand dead bees together with plant samples from the surrounding crops that have been treated with pesticides. If the examination of the samples confirms that the bees have been poisoned, you may be able to resort to legal action against the person responsible.

Colonies damaged in this way require special care and attention. It will normally be necessary to unite colonies to ensure they are big enough. The united stock will revive if they are fed honey.

INTERNATIONAL BEEKEEPING

by Ron Brown, OBE

Beekeeping has an ancient and honourable history over most of the world, with records going back far beyond the ancient civilizations, to cave paintings as far apart as southern Spain and central Africa, dating back to 12,000 years BC.

Old Testament biblical references abound, from the Psalms of David to the account of Samson taking a wild honeycomb to his girl-friend Delilah.

Bees have been kept in hollow logs, bark hives, pottery vessels, willow baskets, straw skeps, even purposely in hollow walls of houses, since pre-historic times.

Hives and Bees the World Over

Predominance of the Langstroth Hive

Today most of the world's honey bees are kept in Langstroth hives, comprising one or more rectangular wooden boxes, each holding eleven deep frames providing about 65,000 cells in a space of about 1½ cubic feet or 40 litres.

In fact, in the great honey-producing countries of North and

A beehive designed for African beekeepers, on display at the Embu Agricultural Show in Kenya

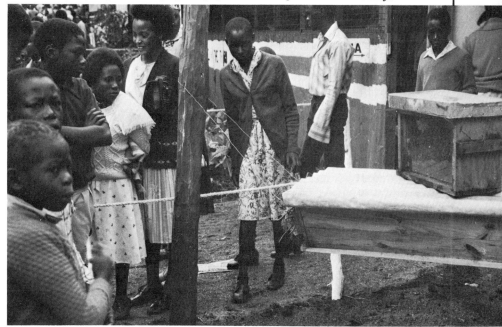

South America, Australia and New Zealand, it is taken for granted that this type of hive is used, so that the actual name is never even mentioned.

Even in China recently I saw mostly apiaries of Langstroths, locally made to standard dimensions, except that rather thicker wood had been used.

I was told at the Institute of Apicultural Research in Beijing (Peking) that currently out of 6 million hives of bees in China, over 4 million were now modern Langstroths housing European *mellifera* bees, with the slightly smaller Asiatic *cerana* bees in other types of hive being phased out.

European Hives

In Europe many colonies are kept in Langstroth hives made to international standards, but each European country still has its own types, some smaller but some larger than the Langstroth.

In Britain almost 90 per cent of all hives are either Nationals or WBCs, both using identical British Standard frames, eleven in a National and ten in a WBC. The so-called British Commercial hive takes eleven (sometimes twelve) rather larger frames, and the Dadant is larger still, being a deeper edition of a Langstroth. In Scotland the Smith hive is very popular, but essentially it is a National adapted to take B.S. frames but with short lugs.

In Ireland there is also the CDB hive, with a deep roof extending down to floor level for greater protection in winter, but Nationals and Commercials are gaining in favour.

Making a Living from Bees

With a very low wage structure and modest material expectations (a bicycle rather than a car, a room rather than a house and electricity only for lighting) a family in China can make a good living from the produce of forty modern hives. In Australia or New Zealand, with comparable honey yields but high wages and high material living standards (one or two cars, a house of five or more rooms and unlimited electricity), it takes 800 similar hives to provide a reasonable living.

In Britain, with our erratic weather pattern, it is almost impossible to make a living from any number of hives, and most so-called bee farmers either pack and sell imported honey as well, deal in hives and bee equipment or have some other income.

In Africa there are still more beehunters than beekeepers, with low yields of honey used locally as a sweetener or to brew a strong wine, and only the beeswax sold. Most of the beeswax of commerce still comes from tropical African countries.

In Singapore earlier this year I found no honeybees at all! With no crops needing pollination and a very large human population living in a small area, bees have been banned and even the Director of the Singapore Science Institute was having great difficulty in gaining permission to instal just one observation hive in his Life Sciences Gallery.

I gave advice on this but I fear to no effect, and Singapore remains the only sovereign state to my knowledge completely without bees.

Even Alaska in the far north manages by having 1.8 kg (4 lb) packages of bees plus young queens flown in mid-April every year from the southern states, to gather a harvest of honey in the short summer before being destroyed in the fall, with combs preserved for the next spring.

World Honey Production

To put our own beekeeping into perspective, a table of world annual honey production for 1986 is given below:

	tonnes per year		tonnes per year
USSR	190,000	Brazil	26,900
China	152,000	Argentina	25,000
USA	91,000	Germany	18,000
Mexico	52,000	France	13,000
Canada	37,800	Japan	5,000
Australia	27,000	UK	3,200

In general terms, the main industrial countries with high standards of living are overall importers of both honey and beeswax, with Europe and Japan the largest. China and Mexico are the largest exporters of honey.

The Varroa Mite

With Varroa now confirmed in the USA (October 1987) and the fact that in 1987 several thousand queen bees plus attendant workers were imported from North America into Britain, it now seems inevitable that this pest is with us, but as yet undetected.

How has it come about that Varroa, for thousands of years confined to the Far East, is now world-wide, except for Australia and New Zealand? Why should it be so much more deadly to our bees than it was to Far Eastern bees?

A successful parasite does not kill its host, and previously for countless thousands of years Varroa was parasitic only on *Apis cerana* (sometimes called *Apis indica*), a slightly smaller cousin of our honeybee, with a somewhat shorter life cycle, existing in India, China and the Far East only.

The shorter life cycle of *cerana* means that Varroa can only develop successfully on drone larvae, which take longer to develop than larvae of workers or queens. Thus Varroa came to take a percentage of drone larvae as a 'tax' on *cerana*, unwelcome perhaps but not lethal, as more drones are produced anyway than a colony normally needs.

This was the compromise evolved by *cerana* and Varroa over many thousands of years, ensuring the survival of both host and parasite.

The World-wide Spread of Varroa

Until the age of world travel, Varroa stayed in the Far East with the *cerana* bees, and no-one elsewhere was interested. Then colonies of *mellifera* (our honey bees) were taken to work alongside colonies of *cerana*, and the pest was transferred from one to the other during the 1930s, and this was almost certainly the first time it ever happened in a bee history of 50 million years or so.

With the longer life cycle of *mellifera* larvae, Varroa was able to propagate successfully in worker cells instead of just in drone cells, and found life so much easier that it became lethal instead of just a nuisance.

By itself this is still only an interesting piece of biology, but modern means of communication (trans-Siberian railway and aircraft) have enabled stocks of bees to be moved around the world, and so Varroa came to Russia from the east by rail and to Europe and South America by air.

High yields of honey in Far Eastern Russia were wrongly attributed to the excellence of bees there and stocks were taken to European Russia (plus Varroa) to 'improve' honey yields. Likewise queens plus workers were imported into central Europe for experimental work, again bringing the pest we need never have had.

At about the same time Varroa came to South America via queens and workers imported into Argentina from Japan, and I personally saw Varroa for the first time in southern Brazil about ten years ago (1978).

The rapid spread of Varroa within a large new territory is also due to modern transportation, especially in North America where thousands of colonies are regularly trucked right across the continent. For example, bees are transported every year in autumn, after completing blueberry pollination in New Hampshire, to winter sites in Florida (mild, no sugar feeding needed).

Thousands of packages of bees are raised each year in the warm deep south of the USA and flown or trucked north in spring, to restock hives where bees have been killed in the previous autumn (after all the honey has been taken off).

Living with Varroa

So what do we do about it? Firstly, early diagnosis. The established techniques are:

1 Placement of cardboard sheets on hive floors in autumn and removal in spring to check hive debris for Varroa mites. This method is made more efficient by the use of wire mesh standing about 1 cm (just over ¼ inch) up from the card to prevent bees from removing the debris themselves.

2 Using tobacco smoke (cigarette ends collected from pubs about 10 p.m!) to dislodge some adult mites, again on to a card insert on the hive floor, within twenty-four hours.

3 In Brazil ten years ago I saw a ruthless but effective method – shaking a small cluster of 50-100 bees into a glass jar half full of petrol and swilling it around, when any Varroa mites can be seen, separated from hosts, especially if the contents are poured through a coffee filter paper and the bees picked off.

4 Since Varroa by habit still prefers to rear its young in large cells, a comb of drone cells put into a hive in April, at the edge of the brood nest, will attract them, and removal two to three weeks later may allow the black spots of developing Varroa to be seen against the clear white background of larvae or pupae.

One of the difficulties is that, by the time damage to bees makes the infestation noticeable, Varroa has probably been there three years, and so has already been passed on to other stocks. An original infestation of perhaps eight to twenty mites is almost impossible to spot in a hive of 40,000 bees; even the following year when numbers are perhaps just a few hundred.

By the third year, mites are present in thousands and before the end of the fourth year the colony is dead.

The drone comb technique is also used in Germany for Varroa control. About three times between April and June a frame of drone comb is inserted at the edge of the brood nest, where a queen would normally expect to find drone comb. Varroa is attracted here and then three weeks later this comb is taken out and destroyed (or placed in a poultry house for larvae to be pecked out and eaten, and the comb re-used).

This does not eliminate Varroa but keeps the infestation down to a low level, especially if medication with some preparation like Folbex VA is also used at appropriate times.

Another technique which I have seen practised in Germany is to make up nuclei with young queens, taken from stocks either free from or with a very low level of infestation, and build them up rapidly by feeding so that the bees outgrow Varroa and give a crop that year and the next before the slower build-up of Varroa takes serious effect.

An interesting possibility is that some races of honey bees may have greater resistance to Varroa and even be able to remove them physically from adult bees. In conversation with Prof. Roger Morse (Cornell University, USA) recently, I learned that the so-called 'killer bees' of Africa appear to be much more resistant to Varroa than our European honey bees are, but these bees (*Apis mellifera adansoni*, sometimes now called *scutellata*) are reckoned a problem in themselves in America.

A further complication is that a virus infection can enter via wounds inflicted by this parasite, and secondary infection can be more lethal than the mite itself.

Killer Bees

Any look at world beekeeping must include something about the so-called 'killer bees of Brazil', or the 'African killer bees' as they are also called.

Up to about thirty years ago the aggressive bees of Central Africa were unknown outside that continent and received little publicity of any kind, except for newspaper reports of what the British Army had to say about them in the days when Kenya was a colony with Mau Mau problems and a large army base just outside Nairobi. At that time I was keeping these African bees in Broken Hill, Lusaka and Ndola in what is now Zambia.

In 1956 Prof. W. E. Kerr visited Angola, Mozambique, Tanganyika, the Belgian Congo and South Africa on a grant from the Rockefeller Foundation, and transported to Brazil 173 queen bees, from Taborah in Tanganyika and Pretoria in South Africa, of which fifty survived the journey and were introduced into colonies of de-queened *mellifera*.

The object was to improve the European dark bees (*Apis mellifera*

mellifera) introduced a century or more previously by the Portuguese, by introducing new blood from a hard-working tropical bee, *Apis mellifera adansoni*, in the expectation that the honey-getting vigour of *adansoni* allied to the gentle behaviour of the Portuguese dark bees would improve honey yields and beekeeping in Brazil generally.

In the event several swarms escaped into the forest and prospered exceedingly, but also retained in full measure the aggressive nature of their African ancestry, also the propensity to swarm frequently and travel long distances as swarms.

How far will they spread?

After a few years of building up in Brazil, these aggressive bees started to spread out over the whole of South America, crossed the Panama Canal in 1984, have now reached Mexico and are shortly expected to arrive in the United States.

It seems to me unlikely that they will establish themselves further north than Texas and the southern states of the USA, any more than they have in the cooler southern area of Argentina. With a much higher threshold working temperature than our own bees, they would be so late building up in a cold English spring that they would not prosper, or even survive here, in my opinion.

How aggressive are they?

It has to be remembered that the progeny of a queen mated in Africa would be wholly African in nature,

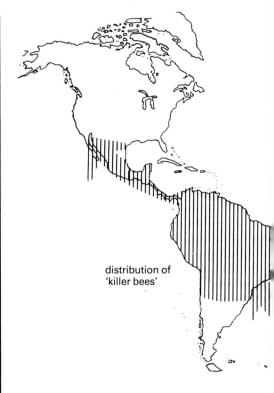

distribution of 'killer bees'

though nurtured by Brazilian bees. The African drones proved to be dominant so far as mating in competition with local bees was concerned, and the 'killer bees' which I saw in Brazil in 1978 were certainly identical in all respects with the African bees I kept in Zambia from 1952-64.

If the bees now colonizing Mexico are indeed hybrids, then from all accounts of their behaviour, the dominant genes must be of African origin.

At least they achieved one purpose in Brazil, that of boosting honey production from 5,000 tons a year in 1956 to five times that quantity thirty years later, but they do need very

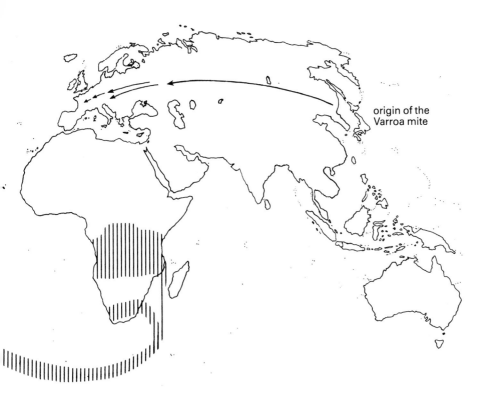

origin of the
Varroa mite

Vertical hatching shows the spread of 'killer bees' into North and South America; the arrows show the route by which Varroa first spread into the west

different management techniques.

Honey production is excellent, and I regularly obtained averages of over 45 kg (100 lb) a year per colony in Central Africa. The same yields are obtained in Brazil.

During the 1960s the press and radio delighted in publishing sensational stories about 'killer bees', although less than fifty people per year were killed in all Brazil (population over 100 millions) by insect stings of any kind, as opposed to 200 people a year in the city of Sao Paulo alone by rabies following dog bites.

Even less publicity was given to more than twenty times this number killed by motor vehicles on the Brazilian roads each year.

Natural aggression, expressed as a readiness to attack and sting, as well as the habit of following victims 100 yards, is real enough. The actual effect of a sting is about the same as that of European bees, no more and no less. I have been stung thousands of times by both kinds of bee and am certain that there is no extra venom in 'killer bees', just many more of them stinging!

ORGANIZATIONS AND SUPPLIERS

compiled by Cecil Tonsley, FRES, Vice-President of Apimondia

Australia

Journals

The Australian Beekeeper
P.M.B. 19 Maitland
N.S.W. 2320

The Australian Bee Journal
Sec. Les Bould
'Willunga', Moonambel
Victoria 3478

Associations

Queen Bee Breeders' Association
Sec. Murray Charlton
Terranora, N.S.W.

Queensland Beekeeper's Association
Sec. Mrs. L.R. Keith
P.O. Box 111, Inglewood
Queensland 4386

South Australian Apiarists' Association
Sec. Mrs. B. Davison
P.O. Mt. Compass, S.A. 5210

Tasmanian Beekeepers' Association
Sec. Mrs. V.N. Ayton
38 James Street, Ulverston
Tasmania 7315

Victorian Apiarists' Association
Sec. Les Bould
'Willunga', Moonambel, Vict. 3478

The Federal Council of Australian
Apiarists' Associations
R.M.B. Glenrowan, Victoria 3675

Manufacturers

John L. Guilfoyle Pty. Ltd.
772 Boundary Road
Darra, Brisbane, Brisbane 4076

Pender Beekeeping Supplies Pty. Ltd.
P.M.B. 19 Gardiner Street
Rutherford, N.S.W. 2320

Tonkins Midland Apiary Suppliers
2 Jocelyn Grove, Springwood
Queensland 4127

Canada

Journal

Canadian Beekeeping
Box 128, Orono
Ontario LOB 1MO

India

Journal

Indian Bee Journal
217 Sadashiv Peth
Pune (Poona) 41130

Association

All India Beekeepers' Association
217 Sadashiv Peth
Pune (Poona) 411030

Ireland

Journal

The Irish Beekeeper (An Beacaire)
Mr. J.J. Doran
St. Judes, Mooncoin, Waterford

Association

Federation of Irish Beekeepers' Associations
Rev. Bro. H.I. Behan
Monkstown Park College
Dunlaoire, Co. Dublin

New Zealand

Journal

The New Zealand Beekeeper
Dalmuir House
The Terrace
P.O. Box 4048, Wellington

The Apiarist
P.O. Box 5056
Papanui, Christchurch

Association

National Beekeepers' Association of
New Zealand
Dalmuir House
The Terrace, P.O. Box 4048, Wellington

Manufacturers

Mahurangi Hiveware
Pukapuka Road
R.D. 3 Warkworth

New Zealand Beeswax Processors Ltd.
Opuha Street, Orari
S. Canterbury

South Africa

Journal

South African Bee Journal
Mr. C.C. Deschodt
P.O. Box 4488, Pretoria 0001

Association

South African Federation of Beekeepers'
  Associations
P.O. Box 4488, Pretoria 0001

United Kingdom

Journals

Bee Craft
15 West Way, Copthorne Bank
Crawley, Sussex

Beekeeping
Clifford Cottage
42a Clifford Street
Chudleigh, Devon, TQ13 0LE

Bee World
18 North Road
Cardiff, Glam. CF1 3DY

Beekeepers' News
Beehive Works, Wragby
Lincoln, LN3 5LA

The Scottish Beekeeper
Firparkneuk, Kirtlebridge
Lockerbie, DG11 3LZ

British Bee Journal
46 Queen Street, Geddington
Kettering, Northants. NN14 1AZ

Associations

British Beekeepers' Association
National Agricultural Centre
Stoneleigh Park
Kenilworth, Warks. CV8 2LZ

Scottish Beekeepers' Association
9 Glenholme Avenue, Dyce
Aberdeen, AB2 0FF

Ulster Beekeepers' Association
Sec. Lawson Swinerton
Moyola Lodge, Castledawson
Co. Derry

Welsh Beekeepers' Association
Sec. Mrs. Pam Gregory
Pentre Bwlen Llandewi Brefi Tregaron
Dyfed Llangibby

International Bee Research Association
18 North Road
Cardiff, CF1 3DY

Manufacturers

Exeter Bee Supplies
64 Haven Road
Exeter, EX2 8DJ

Kemlea Bee Supplies
Starcroft Apiaries, Catsfield
Battle, Sussex

Maisemore Apiaries
Old Road, Maisemore
Gloucester, GL2 8HT

Steele & Brodie (1983) Ltd.
Stevens Drove, Houghton
Stockbridge, Hants. SO20 6LP

E.H. Thorne (Beehives) Ltd.
Beehive Works, Wragby
Lincoln, LN3 5LA

USA

Journals

American Bee Journal
Hamilton, Illinois 62341

Gleanings in Bee Culture
Medina, Ohio, 44258

The Speedy Bee
P.O. Box 998, Jessup, Pa. 31545

Associations

The American Beekeeping Federation Inc.
13637 N.W. 39th Avenue
Gainsville, Florida 32601

Eastern Apicultural Society of
  North America, Inc.
5 Rooney Street
Northborough, Mass. 01532

Manufacturers

The A.I. Root Co, Inc.
P.O. Box 706, Medina, Ohio
Dadant & Sons, Inc.
Hamilton, Illinois 62341

The Walter T. Kelley Co.
Clarkson, Kentucky 42726

Maxant Industries
Dept. G
P.O. Box 454, Ayer, Mass. 01432

York Bee Company
P.O. Box 307
South Macon Street, Ext.
Jessup, Pa. 31545

INDEX

Page numbers in **bold type** refer to pictures